T0214974

Advances in Intelligent Systems and Computing

Volume 567

Series editor

Janusz Kacprzyk, Polish Academy of Sciences, Warsaw, Poland
e-mail: kacprzyk@ibspan.waw.pl

About this Series

The series "Advances in Intelligent Systems and Computing" contains publications on theory, applications, and design methods of Intelligent Systems and Intelligent Computing. Virtually all disciplines such as engineering, natural sciences, computer and information science, ICT, economics, business, e-commerce, environment, healthcare, life science are covered. The list of topics spans all the areas of modern intelligent systems and computing.

The publications within "Advances in Intelligent Systems and Computing" are primarily textbooks and proceedings of important conferences, symposia and congresses. They cover significant recent developments in the field, both of a foundational and applicable character. An important characteristic feature of the series is the short publication time and world-wide distribution. This permits a rapid and broad dissemination of research results.

Advisory Board

Chairman

Nikhil R. Pal, Indian Statistical Institute, Kolkata, India
e-mail: nikhil@isical.ac.in

Members

Rafael Bello Perez, Universidad Central "Marta Abreu" de Las Villas, Santa Clara, Cuba
e-mail: rbellop@uclv.edu.cu

Emilio S. Corchado, University of Salamanca, Salamanca, Spain
e-mail: escorchado@usal.es

Hani Hagras, University of Essex, Colchester, UK
e-mail: hani@essex.ac.uk

László T. Kóczy, Széchenyi István University, Győr, Hungary
e-mail: koczy@sze.hu

Vladik Kreinovich, University of Texas at El Paso, El Paso, USA
e-mail: vladik@utep.edu

Chin-Teng Lin, National Chiao Tung University, Hsinchu, Taiwan
e-mail: ctlin@mail.nctu.edu.tw

Jie Lu, University of Technology, Sydney, Australia
e-mail: Jie.Lu@uts.edu.au

Patricia Melin, Tijuana Institute of Technology, Tijuana, Mexico
e-mail: epmelin@hafsamx.org

Nadia Nedjah, State University of Rio de Janeiro, Rio de Janeiro, Brazil
e-mail: nadia@eng.uerj.br

Ngoc Thanh Nguyen, Wroclaw University of Technology, Wroclaw, Poland
e-mail: Ngoc-Thanh.Nguyen@pwr.edu.pl

Jun Wang, The Chinese University of Hong Kong, Shatin, Hong Kong
e-mail: jwang@mae.cuhk.edu.hk

More information about this series at http://www.springer.com/series/11156

Rituparna Chaki · Khalid Saeed
Agostino Cortesi · Nabendu Chaki
Editors

Advanced Computing and Systems for Security

Volume Three

Springer

Editors
Rituparna Chaki
A.K. Choudhury School of Information
    Technology
University of Calcutta
Kolkata, West Bengal
India

Khalid Saeed
Faculty of Computer Science
Bialystok University of Technology
Białystok
Poland

Agostino Cortesi
DAIS—Università Ca' Foscari
Mestre, Venice
Italy

Nabendu Chaki
Department of Computer Science
    and Engineering
University of Calcutta
Kolkata, West Bengal
India

ISSN 2194-5357   ISSN 2194-5365   (electronic)
Advances in Intelligent Systems and Computing
ISBN 978-981-10-3408-4   ISBN 978-981-10-3409-1   (eBook)
DOI 10.1007/978-981-10-3409-1

Library of Congress Control Number: 2016960183

Printed on acid-free paper

This Springer imprint is published by Springer Nature
The registered company is Springer Nature Singapore Pte Ltd.
The registered company address is: 152 Beach Road, #21-01/04 Gateway East, Singapore 189721, Singapore

Preface

The Third International Doctoral Symposium on Applied Computation and Security Systems (ACSS 2016) took place during August 12–14, 2016 in Kolkata, India organized by University of Calcutta in collaboration with NIT, Patna. The Ca Foscari University of Venice, Italy and Bialystok University of Technology, Poland were the international collaborators for ACSS 2016.

As the symposium turned to its third consecutive year, it has been interesting to note the kind of interests it has created among the aspiring doctoral scholars and the academic fraternity at large. The uniqueness of being a symposium created specifically for Ph.D. scholars and their doctoral research has already established ACSS as an inimitable event in this part of the globe. The expertise of the eminent members of the program committee helped in pointing out the pros and cons of the research works being discussed during the symposium. Even during presentation of each paper, the respective Session Chair(s) had responded to their responsibilities of penning down a feedback for more improvements of the paper. The final version of the papers thus goes through a second level of modification as per the session chair's comments, before being included in this post-conference book.

Team ACSS is always on the look-out for latest research topics, and in a bid to allow such works, we always include them to our previous interest areas. ACSS 2016 had Android Security as the chosen new interest area, thus inviting researchers working in the domains of data analytics, cloud and service management, security systems, high performance computing, algorithms, image processing, pattern recognition.

The sincere effort of the program committee members, coupled with indexing initiatives from Springer, has drawn a large number of high-quality submissions from scholars all over India and abroad. A thorough double-blind review process has been carried out by the PC members and by external reviewers. While reviewing the papers, reviewers mainly looked at the novelty of the contributions, at the technical content, at the organization and at the clarity of presentation. The entire process of paper submission, review and acceptance process was done electronically. Due to the sincere efforts of the Program Committee and the Organizing Committee members, the symposium resulted in a suite of strong technical paper presentations followed by effective discussions and suggestions for improvement for each researcher.

The Technical Program Committee for the symposium selected only 21 papers for publication out of 64 submissions after peer review. Later 5 more papers were chosen for presenting in the symposium after the authors submitted enhanced versions and those were reviewed again in a 2-tier pre-symposium review process.

We would like to take this opportunity to thank all the members of the Program Committee and the external reviewers for their excellent and time-bound review works. We thank all the sponsors who have come forward towards organization of this symposium. These include Tata Consultancy Services (TCS), Springer Nature, ACM India, M/s Neitec, Poland, and M/s Fujitsu, Inc., India. We appreciate the initiative and support from Mr. Aninda Bose and his colleagues in Springer Nature for their strong support towards publishing this post-symposium book in the series "Advances in Intelligent Systems and Computing". Last but not least, we thank all the authors without whom the symposium would have not reached this standard.

On behalf of the editorial team of ACSS 2016, we sincerely hope that ACSS 2016 and the works discussed in the symposium will be beneficial to all its readers and motivate them towards even better works.

Kolkata, India   Rituparna Chaki
Białystok, Poland   Khalid Saeed
Mestre, Venice, Italy   Agostino Cortesi
Kolkata, India   Nabendu Chaki

About the Book

This book contains extended version of selected works that have been discussed and presented in the Third International Doctoral Symposium on Applied Computation and Security Systems (ACSS 2016) held in Kolkata, India during August 12–14, 2016. The symposium was organized by the Departments of Computer Science and Engineering and A.K. Choudhury School of Information Technology, both from University of Calcutta in collaboration with NIT, Patna. The International partners for ACSS 2016 had been Ca Foscari University of Venice, Italy and Bialystok University of Technology, Poland.

This bi-volume book has a total of 26 chapters divided into 7 chapters. The chapters reflect the sessions in which the works have been discussed during the symposium. The different chapters in the book include works on data analytics, algorithms, high performance computing, cloud and service management, image processing, pattern recognition, and security systems.

The effort of the editors and all the contributing authors will be meaningful if the books are used by the contemporary and future researchers in the field of computing.

Contents

About the Editors

Rituparna Chaki is Professor of Information Technology in the University of Calcutta, India. She received her Ph.D. Degree from Jadavpur University in India in 2003. Before this she completed B.Tech. and M.Tech. in Computer Science and Engineering from University of Calcutta in 1995 and 1997, respectively. She has served as a System Executive in the Ministry of Steel, Government of India for 9 years, before joining the academics in 2005 as a Reader of Computer Science and Engineering in the West Bengal University of Technology, India. She has been with the University of Calcutta since 2013.

Her areas of research include optical networks, sensor networks, mobile ad hoc networks, Internet of Things, data mining, etc. She has nearly 100 publications to her credit. Dr. Chaki has also served in the program committees of different international conferences. She has been a regular Visiting Professor in the AGH University of Science and Technology, Poland for last few years. Rituparna has co-authored couple of books published by CRC Press, USA.

Khalid Saeed is Full Professor in the Faculty of Computer Science, Bialystok University of Technology, Bialystok, Poland. He received the B.Sc. degree in Electrical and Electronics Engineering in 1976 from Baghdad University in 1976, the M.Sc. and Ph.D. degrees from Wroclaw University of Technology, in Poland in 1978 and 1981, respectively. He received his D.Sc. degree (Habilitation) in Computer Science from Polish Academy of Sciences in Warsaw in 2007. He was a visiting professor of Computer Science with Bialystok University of Technology, where he is now working as a full professor. He was with AGH University of Science and Technology during 2008–2014. He is also working as Professor with the Faculty of Mathematics and Information Sciences in Warsaw University of Technology. His areas of interest include biometrics, image analysis and processing and computer information Systems. He has published more than 220 publications, edited 28 books, journals and conference proceedings, 10 text and reference books. He has supervised more than 130 M.Sc. and 16 Ph.D. theses. He gave more than 40 invited lectures and keynotes in different conferences and universities in Europe, China, India, South Korea and Japan on biometrics, image analysis and processing. He has received more than 20 academic awards. Khalid Saeed is a member of more than 20 editorial boards of international journals and conferences. He is an IEEE Senior Member and has been selected as IEEE Distinguished Speaker for 2011–2016. Khalid Saeed is the Editor-in-Chief of International Journal of Biometrics with Inderscience Publishers.

Agostino Cortesi, Ph.D., is Full Professor of Computer Science at Ca' Foscari University, Venice, Italy. He served as Dean of the Computer Science studies, as Department Chair, and as Vice-Rector for quality assessment and institutional affairs.

His main research interests concern programming languages theory, software engineering, and static analysis techniques, with particular emphasis on security applications. He published more than 110 papers in high-level international journals and proceedings of international conferences. His h-index is

16 according to Scopus, and 24 according to Google Scholar. Agostino served several times as member (or chair) of program committees of international conferences (e.g., SAS, VMCAI, CSF, CISIM, ACM SAC) and he's in the editorial boards of the journals "Computer Languages, Systems and Structures" and "Journal of Universal Computer Science". Currently, he holds the chairs of "Software Engineering", "Program Analysis and Verification", "Computer Networks and Information Systems" and "Data Programming".

Nabendu Chaki is Professor in the Department of Computer Science and Engineering, University of Calcutta, Kolkata, India. Dr. Chaki did his first graduation in Physics from the legendary Presidency College in Kolkata and then in Computer Science and Engineering from the University of Calcutta. He completed Ph.D. in 2000 from Jadavpur University, India. He shares six international patents including four U.S. patents with his students. Professor Chaki has been quite active in developing international standards for software engineering and cloud computing as a member of Global Directory (GD) member for ISO-IEC. Besides editing more than 25 book volumes, Nabendu has authored six text and research books and has more than 150 Scopus Indexed research papers in journals and international conferences. His areas of research interests include distributed systems, image processing and software engineering. Dr. Chaki has served as a Research Faculty in the Ph.D. program in Software Engineering in U.S. Naval Postgraduate School, Monterey, CA. He is a visiting faculty member for many universities in India and abroad. Besides being in the editorial board for several international journals, he has also served in the committees of over 50 international conferences. Professor Chaki is the founder Chair of ACM Professional Chapter in Kolkata.

Part I
Algorithms

A Heuristic Framework for Priority Based Nurse Scheduling

Paramita Sarkar, Rituparna Chaki and Ditipriya Sinha

Abstract Nurse Scheduling Problem is traditionally studied as a multi objective problem which aims at an optimum scheduling of nurse assignment to patients in a hospital. Remote healthcare is involved with providing quality care and medical assistance to patients based on remote monitoring. This paper presents an in-depth study of nurse scheduling algorithms. This study is followed with a description of the existing logics and the open issues. It also presents a heuristic framework with new features and a variable wait time based multi-parametric cost function for managing the dynamic NSP associated to remote healthcare.

Keywords Remote healthcare · Nurse scheduling problem · Constraints · Dynamic · Priority of patients · Heuristic · Cost · Shuffling

1 Introduction

Healthcare systems are expanding their services in various domains like remote patient monitoring, e-care services, telemedicine services, assistive living services, etc. These services involve patient monitoring as an intrinsic part in all healthcare systems. Different types of sensing devices, networking services, and servers are being deployed and incorporated in the remote healthcare systems to provide

P. Sarkar (✉)
Calcutta Institute of Engineering and Management,
Tollygunge, Kolkata, India
e-mail: mailtoparo@gmail.com

R. Chaki
A.K. Choudhury School of Information Technology,
University of Calcutta, Kolkata, India
e-mail: rituchaki@gmail.com

D. Sinha
National Institute of Technology (NIT—PATNA), Patna, India
e-mail: ditipriyasinha87@gmail.com

© Springer Nature Singapore Pte Ltd. 2017
R. Chaki et al. (eds.), *Advanced Computing and Systems for Security*,
Advances in Intelligent Systems and Computing 567,
DOI 10.1007/978-981-10-3409-1_1

patients a faster and efficient response from the medical staff. This involves accurate and quick decision making, proper and effective allotment of caregivers, i.e., nurses, and assistance from a doctor at regular time interval. Scheduling of nurses thus becomes a complex decision making involving a nurse and a patient subject to a variety of constraints [1–4]. The need of staff scheduling and rostering is evident from researches. The efficient schedule is useful in building timetable which again can be reused to make replacing, shuffling, etc., in the next week schedule. Absenteeism [5], demands, shuffling, etc. are the related issues that are also considered for efficient scheduling. The remote nurse scheduling is more complicated as it involves continuous monitoring of patients. The challenges of remote nurse scheduling lie in quick decision making and assignment of graded nurse in complex environment, efficient time management, classification of structured and unstructured heterogeneous data, prioritization of patients, etc.

In Sect. 2 a brief overview of the modelling of NSP and related application is described in bibliographic survey Sect. 2, in Sect. 3, the open issues have been discussed. In Sect. 4, the proposed model of Remote Nurse Scheduling is presented.

2 Bibliographic Survey

The following section presents the different directions to solve Nurse Scheduling Problem. Previous researches in this domain have revealed that Nurse Scheduling Problem (NSP) is a multi-objective problem [1, 6] and nurse work schedule has an independent effect on patients outcome [7]. Systematic approach for nurse allocation is needed to ensure continuous and adequate level of patient care services while maintaining the legislative requirements as well as internal policies. A number of mathematical models [8–11] were formulated to implement optimized solution for NSP.

2.1 Integer Programming Model

In nurse scheduling problem, nurses' preference or demand, and health centres' restrictions are considered for making better assignment. Many efficient algorithms with various mathematical formulations and techniques had been proposed and evaluated in earlier years to solve this problem. In remote Healthcare scenario, patients who are situated at remote places from the hospital may need continuous monitoring services and advices from the caregivers on emergencies. Therefore, at the time of making of a schedule a large number of constraints are needed to be considered as hard as well soft constraints [2].

Prior to the development of mathematical programming, most nursing scheduling approaches were based on cyclical modelling [12]. As one of the major problems of nurse scheduling problem is the potential size of the set of individual schedules, highly repetitive and rigid cyclic schedules are not suitable in supply of and demand for nursing services. In this regards, several studies come with various information. The most solutions using the integer or mixed linear programming approach [13–15] had been used to solve the nurse scheduling problem by involving a linear relaxation of the master problem. A series of multi-objective integer programming models [1, 6] are formulated for the nurse scheduling problem where both nurse shift preferences and patient workload as for patient dissatisfaction are considered in the models. A two-stage non-weighted goal programming solution method is used to find an efficient solution to solve the nurse schedule for a specified horizon of time in a way that satisfies both hard and soft constraints whereas other instances use Integer Programming model to build nurse roistering and scheduling [1] with real world constraints. This paper [1] uses optimization logic to find a solution with the minimum number of violations on the coverage-related hard constraint in a faster time by assigning one of the available shift patterns to each nurse. Earlier work [11] proposes an Integer Quadratic programming technique to minimize "shortage cost" of nursing care services while satisfying various assignment constraints. Here nurses' preference costs are not considered to build feasible solution, on the contrary, [14] uses column generation techniques where the columns correspond to individual schedules for each nurse.

Since remote healthcare requires dynamic decision making, nurse scheduling needs to be formulated as an optimization problem to tackle the complex needs of the society. In this view, a linear integer programming model [10] is developed for constructing the schedules of University of Pittsburgh Medical Center (UPMC). Here schedules are optimized considering both the soft constraints and hard constraints. Objective function has been built to minimize the cost including nurse's aversion and hospital's soft constraints so that fitness value of the feasible shifts gives minimum cost and thus better schedule. However, this work does not consider limiting the infeasible solutions whereas a method of ranking algorithms [16] is developed on a problem instance that is applicable when some solutions are infeasible by using an integer programming model is used to obtain the fitness value of nurse's shift pattern. Another integer programming model [17] is formulated to solve the nurse scheduling problem (NSP) which satisfies nurse preferences for a practical application of a real-life hospital. Here a complex model is developed to build the schedules of weekdays and weekends with different requirements and different preferences for employees. The model can efficiently satisfy the set of requirements in a particular real life application. However, it fails to generate an adaptive solution, which is taken care of in [18] using mixed integer programming model. The authors deploy a general care ward to replace and automate the current manual approach for scheduling. The developed model differs from other similar studies in that it optimizes both hospital requirement as well as nurse preferences by allowing flexibility in the transfer of nurses from different duties. These adaptable solutions did not consider priority as one of the metrics influencing scheduling. In contrast, remote patients are prioritized [3] based on their heterogeneous health data and multi-objective functions are presented to solve the conflicts and cost of NSP.

2.2 Satisfiablity Model

Nurse Scheduling Problems in the remote healthcare scenario has to consider a variety of parameters for efficient real time decision support. In the literature, various satisfiablity models are studied for solution of NSP. Constraint satisfaction is observed as an important tool to generate an adaptive solution for nurse scheduling [18]. Here the authors take care of the results from the allocation of early duties and late duties in those schedules by incorporating the tabu search method to formulate the more complex problem of allocating each nurse to a pattern of days or nights. Earlier studies reveal that many of the NSP solution methods are based on Branch and Bound Method [19], but they are more time consuming. Paper [20] uses Constraint Programming (CP) to solve the scheduling problems in a French hospital. The model is based on a Constraint Programming language, consisting of a certain level of constraints to be satisfied. The approach is able to produce satisfactory schedules over a planning horizon up to six weeks. In [21], authors present a CP method for solving a week long NRP in a Hong Kong hospital. Here a redundant modelling idea is described in which, both formulations of shifting and assigning are simultaneously updated and fed back into each other. The final result to the problems is presented by the percentage of the satisfaction of soft constraints. But this model does not consider the global soft constraints. On contrary, a satisfiabililty model [4] presents a hybrid approach which emphasizes the application of soft global constraints. The interaction among the global constraints is investigated through the communication among them. None of the above models consider the reduction of the search space. In contrast, authors [22] apply Contrast Programming to solve the Nurse Rostering Problem (NRP) with the help of meta-level reasoning and probability-based order heuristic. The meta-level reasoning is executed before the search to generate redundant or implied constraints from the existing constraints. These new constraints can help in further reducing the search space. It approximates the probability of value assuagements occurring in the solution set and thus uses this information to guide the search. Although many constraint satisfiability models are proposed and built up but none of them did not provide any flexible approach. In this regards, authors [23] provide an optimized technique for feasibility testing among existing constraint solvers by using a flexible weight-age for breakage of soft constraints. A stochastic approach is proposed for integration of the solvers and configuration of distributed multi-agent system in future. One recent approach is discussed [24] for many real-world applications that involve planning and scheduling to assign values to multiple variables in order to satisfy all required constraints. To optimize the scheduling cost penalty functions are proposed in order to balance the objective [25] by using stochastic ranking. The results show that the stochastic ranking method is better in finding feasible solutions but fails to obtain good results with regard to the objective function in [26].

2.3 Meta-Heuristic Search Techniques

Nurse staffing and scheduling are the long term planning that determines the required number of nurses of different qualification levels. As long computation time and memory usage are involved in NSP, a common method is required to be applied. Therefore some kind of heuristic approach [27, 28] is needed to find a solution close to the optimal one. As NSP is NP-hard, meta-heuristic approach is required to solve the problem heuristically in reasonable time. Several meta-heuristic methods [29, 30] are applied to solve NSP in literature. A few of them has been reviewed as follows.

Genetic Algorithm (GA) Based Approaches.

Genetic Algorithm is an evolutionary algorithm which is used to handle the increased number of objectives with ease. GA has become increasingly popular for solving complex optimisation problems such as those found in the areas of scheduling or timetabling. The state of the art study reveals that some instances use genetic algorithm (GA) to minimize the hospital expenses as the work schedules are able to reduce the waiting time of the number of patients. Some of the solutions work with a cost bit matrix for genetic operators to improve performance of GA for NSP, using the selection and crossover operators based on the probability only. A genetic algorithm approach is described in [8] to construct the scheduling problem in a UK based hospital. It uses an indirect coding based on permutations of the nurses, and a heuristic decoder that builds schedules from these permutations. To improve the performance, [26] proposes a hybridized hyper-heuristic (SAHH) algorithm within a local search and genetic algorithm framework to find feasible solution in reduced search space. But the classical programming model has the limitation due to the conflicts rises from the large number of restrictions and time bound complexities. To overcome this problem, a genetic algorithm for nurse scheduling is proposed in [31] using two-point crossover and random mutation to minimize the hospital expenses as the work schedules from proposed method are able to reduce the waiting time of the number of patients from 34.08% to less than 15%. A constructive heuristic is described [32] based on specific encoding and genetic operators for sequencing the scheduling problems which are applied to the nurse re-rostering problem. Here, the permutations of nurse and their tasks are formed which is fed into the heuristic function to re-assign them in order to build feasible schedules. In this work cost function is not applied on genetic operator. A cost bit matrix for genetic operators [33] is used to improve performance of GA for NSP using selection and crossover operators based on the probability only. On the contrary, the mutation operator was applied based on the probability and the value in the cost bit matrix. This approach results in pruning of search space following the reduction in execution time. Authors in [34] develop a two-stage mathematical model for a nurse scheduling system for various applications like hospital management requirements, government regulations, and nursing staffs' shift preferences. Here an empirical case study is performed and the results show that GA can be an efficient tool for solving the nurse scheduling problem. However, as the classical GA has premature convergence to reach at global optimization, [35] uses a new memetic algorithm where local search is applied to each individual selected after selection operation. It produces more optimal results and balances exploitation and exploration within the search space.

Swarm Intelligence Approaches.

To provide the better comparative analysis Particle Swarm Optimization is used [36] to search for minimum unfair assignments among the nurses. This paper lack the consideration of patient-centric service which is addressed in a flexible and adaptive hybrid evolutionary search method based on bee colony optimization [37].

Simulated Annealing Based Approaches.

In meta-heuristic methods some local search techniques are also applied in NSP to construct the adaptive schedules. [38] proposes a mathematical method for nurse scheduling to optimize the hospital limitations and nurse preferences while minimizing the patients' waiting queue time. In this paper authors applied simulated annealing algorithm to obtain the objective in the emergency department of a governmental hospital and to find appropriate schedule. Here the results show that waiting time of achieved schedule from proposed algorithm is 18% less than those from existing schedule. This work lacks the adaptive approach in its proposed method. In nurse scheduling problem optimization is also achieved by reducing the search space in the robust optimized solution. Therefore Cost functions [39] are developed based on the aversion of duties by the nurses. The hospital's constraints are also minimized to formulate the objective function. In this paper authors propose different cost functions to develop a new schedule by applying their transition rule adopting simulated annealing algorithm. Simulated annealing based optimization methods are proposed in [30, 40] to produce an acceptable good solution rather than an optimal solution.

3 Open Issues

From the study of earlier literature review, Nurse scheduling in Remote Healthcare appears to be the next step in the evolution of better healthcare services. However, there are still a number of challenges and open issues that should be faced by the re-search community in order to mature this application. This section exposes the main research directions that will help to create this application. Here we focus on some functional and non-functional requirements that constitute an essential part of the nurse scheduling in remote healthcare as follows:-

Nurse scheduling problem in remote healthcare belongs to a larger domain of appointment scheduling and timetable scheduling problem. It includes very complex, highly constrained environment.

3.1 Identification of the Parameters to be Used for Long-Term Scheduling of Nurses

Analysis of Feasibility. The previous studies on feasibility analysis of shift pattern for assignment of nurses to patients are performed in a fixed setup. But in continuous monitoring of patients in remote healthcare services, the challenges are in proper identification of feasible nurses that could be complex and based on the patients' severity.

Satisfy Large Number of Constraints. Nurses assigned in variable time duration based on the demand of patients, certain restrictions and availability of skilled nurses. Nurse scheduling for a long hour depends on the patient's urgency, nurse's fatigue, job satisfaction and many other parameters. Therefore there are research challenges for implementation of proper methods to recognize those parameters, constraints and searching method for appropriate nursing resource in the changing environment.

3.2 Decision Support System (DSS) in Making Dynamic or Flexible Scheduling

An effective Medical Decision Support System (MDSS) is required to be developed to support the efficient and timely assignment of nurses in real time. The following objectives are specified to be answered by the existing works in this field.

Dynamic Data Mining Algorithm. As Physicians, nurses and other healthcare professionals use a MDSS to prepare a faster and appropriate diagnosis, so the challenge lies how to implement the optimized data mining algorithms in remote nurse scheduling for discovering patterns in medical data of remote patients, association of patterns with unknown events, classification, and prediction of patients' urgency. Based on the efficient data mining algorithms, the assignment of nurses may be conducted in a dynamic environment for remote patient monitoring to examine the patient's medical history in conjunction with relevant clinical research. Such analysis can help to predict potential events, which can range from drug interactions to disease symptoms.

Context-Aware Knowledge Base System. We need an extensible platform for a decision support system to assign suitable nurses to critical patients with an accurate, robust inference engine and knowledge base system. Those systems should incorporate context aware rules which can extract meaningful features from the context for identifying patients' current state, nurses' performances, etc. Without a context-aware knowledge base, on the other hand, we cann't be able to analyze clinical data.

Data Integration. The foremost challenge is that a MDSS must be integrated with a healthcare organization's clinical workflow, which is often already complex. Some clinical decision support systems are standalone products that lack interoperability with reporting and electronic health record (EHR) software. The sheer number of clinical research and medical trials being published on an ongoing basis makes it difficult to incorporate the resulting data. Furthermore, large amounts of heterogeneous data are incorporated into existing systems; therefore significant strain is required on application of data integration tools. To exchange, share, and index various data in heterogeneous formatting, it requires a uniform data integration tool to incorporate data portability.

4 Description of the Heuristic Method for Nurse Assignment in the Remote Centre

Objective. Keeping in mind the above issues, we have proposed an efficient assignment of nurse to a remote patient, based on patient need, nurse expertise, and a variety of parameters, within a pre-fixed time interval. The cost function we have proposed here is a multi-parametric function which is implemented as the fitness function in finding the best feasible solution as shown in Sect. 4.3.

Salient New Features of the Proposed Method. In the literature, various programming model are studied for solution of NSP. One of the major problems of nurse scheduling is the potential size of the set of individual schedules due to large number of parameters and constraints. There are lots of classical as well as hybrid methods have been studied in his domain and many algorithms have been devised in making optimised shift patterns in schedules. But appropriate utilization of nurses according to patients' demand and quick evaluation of timing parameters of the assignment in emergency condition are still absent in those methods.

Therefore some new considerations are proposed and used in this research paper to implement dynamic nurse scheduling. They are as follows:

- A generic multi-parametric nurse assignment method is devised in this paper which implements the cost function on basis of evaluation of a proposed mapping function. The mapping function assigns appropriate nurse as per her expertise, grades to multiple patients according to patients' changing health conditions. It can be useful for dynamic decision making in future.
- An utilization factor of scheduled patients and availability factor of graded nurses are attained here for periodic scheduling with varying waiting timing constraint to minimize the number of patients waiting in the wait queue on basis of which total cost function is measured.
- In this paper the schedule of nurse assignment is shuffled after a period of time on the basis of previous factors for real-time monitoring in order to accommodate the unscheduled patients to each nurse.

Assumption:

- A nurse will be assigned to a patient only if the respective feasibility indices match.

- Free Nurses are sorted in the available pool according to their higher grades scaled as 1 to 4.
- Patients are sorted in the list according to their higher priority which is scaled as 1 to 4 (urgent, high, medium, low).

Hard Constraints. The hard constraints are assumed so that the scheduling algorithm do not violet them.

HC1. No nurse is assigned to work a night shift followed immediately by a day shift

HC2. Higher priority patients should replace lower priority patients in each shift type.

HC3. Minimum number of remote patient in one shift type is one.

HC4. Minimum number of hours for one assigned remote patient is $T_{serv} = 5$.

HC5. Demand of a higher graded nurse is higher than lower graded nurse in all shift types.

HC6. Highest graded nurse should be assigned to highest priority patient in all shift types.

HC7. Each feasible nurse N_i should be assigned to maximum 5 remote patients in a day ($N_{i\_TOTP} = 5$)

Decision Parameters.

Feasible nurse N_f.: A nurse N is called Feasible Nurse N_f if her assignment a_{ijk} satisfies all the 7 hard constraints (HC1 to HC7).

Feasible patient P_f.: A patient is called Feasible Patient P_f if that patient's priority matches with the corresponding Grade G_p of nurse N.

Variable X_{ips}.: It is defined as a decision variable for assignment of a nurse N. X_{ips} is 1 if ith nurse is assigned to a patient p in sth shift type otherwise 0.

Feasible Nurse to Feasible Patient Mapping Function FNP(): FNP() is defined as a mapping between a feasible nurse N_f to a feasible patient P_f. the value is 1 if one feasible nurse N_f is assigned to a feasible patient P_f from her list, otherwise 0.

$$FNP\{0, 1\}: N_f \to P_f \tag{1}$$

Cost function C of a feasible nurse N_f.(CN_f): It is defined as a function where decision parameters are described in [2] to minimize the sum of all penalty cost P_{ij} and demand R_{ijk} or preference cost of all nurses covering all shift patterns of one full week:

$$\sum_{i=1}^{n} \sum_{j=1}^{2} p_{ij} X_{ips} + \sum_{s \in F(i)} \sum_{k=1}^{7} \min \left[\sum_{i=1}^{n} \sum_{j=1}^{2} |(X_{ips} t_{ie} a_{ijk}) - R_{ijk}|; 0 \right] \tag{2}$$

4.1 Methodology

The following methodology is proposed to implement

- a systematic approach for a dynamic assignment procedure based on nurses' expertise and patients' condition
- a variable wait time based cost effective approach
- a balance between the assigned and unassigned patients in the waiting list in terms of utilization factor

The procedure starts with the making of an ordered list AN[i] of available nurses. It is constructed using a greedy search for the minimum cost function. The free nurse N with minimum cost CN_f (Eq. 5) is indexed as N_{min}. This N_{min} is added to AN[i]. Thereafter a prioritized list J of remote patients $J[P_1, P_2,...]$ is made.

1. A matrix NP_feasible [N, P] of N nurses and P remote patients is constructed which consists of mapping function *FNP()* for feasible nurses (N_f) from AN[i] to all feasible patients (P_f) from the list J.
2. N_i represents the nurse to be assigned to P_f according to the NP_feasible [N_f,P_f]. Assign N_i to 1st scheduled P_1 as F: $N_i - > P_1 = 1$.
3. At this stage, a check is made to find the remaining patient J_P in the list J is unscheduled or not for any feasible nurse N_i. If it returns a 'YES', then

    (a) For each nurse N_i in the NP_feasible [N_f,P_f] calculate total duty hours T_{TOT} for all assigned patients.
    (b) Thus the lowest duty hours T_{MIN} is found for a nurse N_{MIN} if $T_{MIN\_TOTP} < 5$ (Sect. 4, HC7), then the unscheduled patient J_P is added to the feasible list of nurse N_i.
    (c) To accommodate each unscheduled patient J_p in the schedule, the schedule is shuffled according to checking all ordered N_i and

    if p_{ij} of J_p is "urgent", then {$P_{j+1} = P_j$ and $P_j = J_p$, $P_{j+1} = P_{j+2}$ and so on} for all assigned remote patients for each N_i.
    Then Add J_p to NP_feasible[N_i, P_j] and
    Set F: $N_i - > P_j$ to 1, add P_5 to wait_list [N_i, P_5],
    Set wait timer $t_{Pf} = 0$ for P_{j+1}} $t_{Pf} = t_{Pf} + 1$. Last assigned patient will be added to the wait_list of N_i.

    (d) Otherwise the unscheduled J_p is assigned to the feasible nurse N_f replacing the patient for whom the assignment function FNP() is currently 0. Wait timer t_{Pf} for replaced patient P_i is incremented.

4. Otherwise it is retained in the future list (wait_list) until the availability of the scheduled nurse. The total wait time T_{wait} as per Eq. (4) for a patient P_f is calculated by using the utilization factor (U) as per the following Eq. (3). If it reaches the maximum wait time max_T_{wait}, then patient P_f is reassigned to nurse N_i, otherwise it goes to check for unscheduled J_P for N_i.

$$U = J_{WA}/J_{TOT} \tag{3}$$

$$T_{wait} = \left(t_{pf} \times U\right) \tag{4}$$

where J_{WA} = total number of scheduled patients waiting in the wait_list

J_{TOT} = total number of unscheduled patients in wait_list

5. The Cost Function for all nurses C_{TOT} is recalculated by adding t_{pf} to cost CN_f as in Eq. 5.
6. C_{TOT} is the required objective function which is to be minimized. It is implemented as the fitness function by using classical genetic algorithm approach as fitness function later in this paper to obtain the best value of the fitness function at which the feasible values of the attributes are obtained to build a feasible schedule.

$$C_{TOT} = \sum_{i=1}^{n} \sum_{j=1}^{2} p_{ij}X_{is} + T_{wait} + \sum_{s \in F(i)} \sum_{k=1}^{7} \min\left[\sum_{i=1}^{n} \sum_{j=1}^{2} \left|(X_{is}t_{ie}a_{ijk}) - R_{ijk}\right|; 0\right] \tag{5}$$

Lemma 1 *The maximum wait time of a scheduled urgent patient P_f waiting in wait_list is $max\_T_{wait} > 1$ depending on availability of graded nurse N_f and patient's utilization factor.*

Proof It is seen from Definition 2 and the hard constraint HC6 that, priority Pr_p of a patient is directly proportional to the nurse's expertise or grade G_p. Therefore, $Pr_p \infty G_p, \therefore Pr_p = A \times G_p, A = Pr_p/G_p$. where A is the ratio of priority of patient and available grades of the nurse, A is here termed as availability factor of suitable graded nurse. If prioritized patients in wait_list are assigned to equivalent and appropriate graded nurse, then A = 1. Otherwise, if higher graded nurses is affluent and assigned to any lower priority patients other than urgent, then it can be seen that $A > 1$, otherwise $A < 1$ Now, let H_n be the occurance of urgent priority patients in a period of time T_i. The total wait time T_{wait} is directly proportional to the probability of occurance of urgent patient i.e. H_n. Therefore if the probability of H_n is $P(H_n)$, then, $T_{wait} \infty P(H_n)$ when, availability factor $A < 1$.

$$\therefore T_{wait} = U \times P(H_n)$$

where, U is the utilization factor for patients in the wait_list at that period of time T_i.

Let H1, H2, ... be a sequence of events H_n in some probability space with n = 1, 2, 3, ... Then according to Borel–Cantelli lemma [41], if the sum of the probabilities of the H_n is finite i.e. if $\sum_{n=1}^{\infty} P(H_n) < \infty$, then the probability that infinitely many of them occur is 0. Suppose (H_n) is a sequence of random numbers of occurring urgent patients with $P(H_n = 0)$ for each n. Hence, according to Borel–Cantelli lemma, the probability of $H_n = 0$ occurring for infinitely many n is 0 and probability of H_n is nonzero with probability 1, for all but finitely many n. So, for finite number of occurrence H_n of urgent patients, the total wait time will be

$T_{wait} = U \times 1$ as $P(H_n) = 1$ for finite number of events. Since maximum wait time max_T_{wait} depends on the number of higher graded nurses and number of urgent patients waiting in the list. If number of higher graded nurses are insufficient, i.e. $A < 1$ and if the number of scheduled urgent patients (J_{WA}) is higher than unscheduled patients (J_{TOT}) in wait_list in emergency situation i.e. $J_{WA} > J_{TOT}$, then $U > 1$. Thus, for finite $P(H_n)$, $T_{wait} > 1$ when $A < 1$ and $J_{WA} > J_{TOT}$ in emergency situation. Otherwise for probability of infinite number of events, $T_{wait} = U \times 0 = 0$ as occurring for infinitely many n is 0.

Therefore from the above deduction, we could say that maximum wait time of an urgent patient in the wait_list i.e. max_$T_{wait} > 1$ for a finite number of urgent priority patients to be assigned depending on the availability factor and utilization factor.

4.2 Flow Diagram of the Proposed Framework

The following workflow of the proposed framework for the remote nurse scheduling in this paper has been designed to represent the methodology of the work (Fig. 1) below.

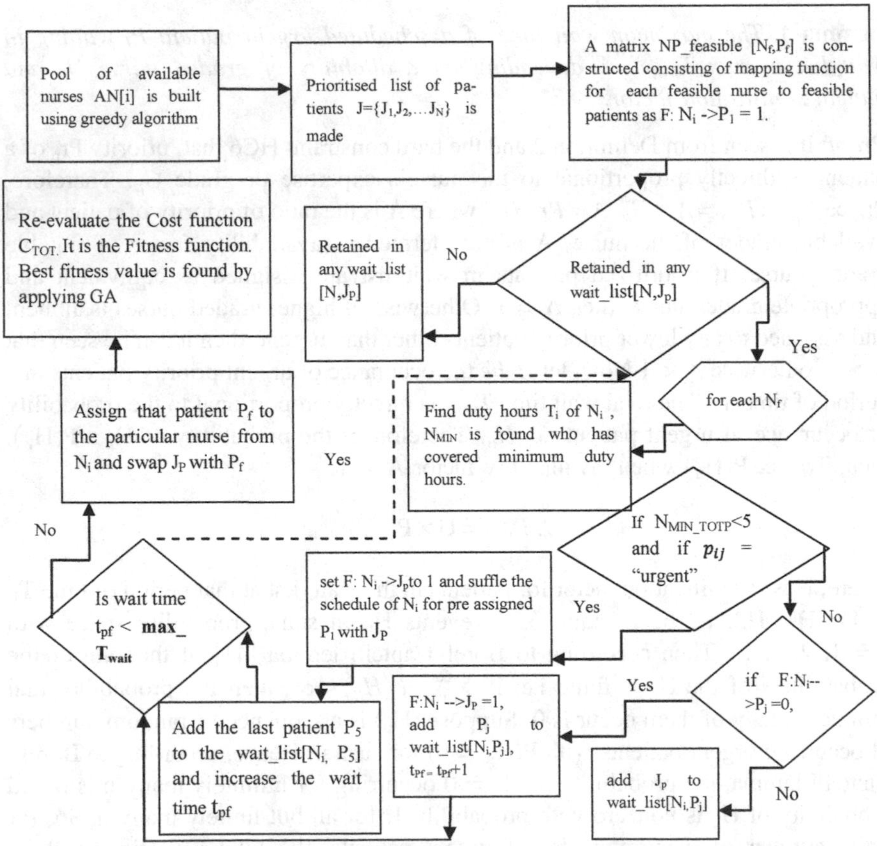

Fig. 1 Flow diagram of the proposed heuristic method for remote nurse scheduling

4.3 Experimental Results

The proposed methodology is implemented in Matlab 7.7.0. In this section the experimented results have been shown. The proposed methodology has been implemented by using Genetic Algorithm Toolbox ("optimtool"). The total cost of the proposed method is considered as the fitness function C_{TOT} which has been implemented here with initial population size 20, with creation function as feasible population type, as well as selection function as Stochastic uniform. The single point crossover and adaptive feasible mutation is used which randomly generates directions that are adaptive with respect to the last successful or unsuccessful generation. The following plot of Best Fitness value vs. number of about 50 generations shows that by applying the above criteria best fitness value after 50 generation is quite minimized as 5 in Fig. 2. It is based on the objective function Eq. (5). In fact, the total wait times in seconds consumed by the 50 generations of 4 instances in GA tool in Matlab 7.7.0 are shown (Table 1) where the number of feasible nurses based on mapping function we developed has found 14 from the collected data of Sri Aurobindo Seva Kendra (EEDF), Kolkata, India. The wait time T_{wait} varies as per the number of waiting patients J_{WA} which determines the patients' utilization factor. The best fitness function which is actually the cost function is obtained here 5 at wait time 2.25 when $J_{WA} = 5$ in the waitlist.

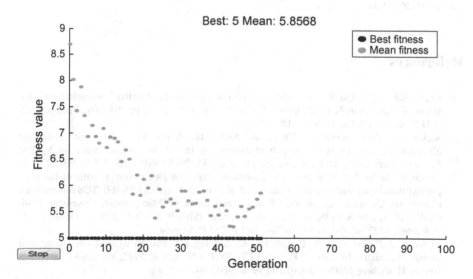

Fig. 2 Plot of fitness value over generations in Matlab 7.7.0

Table 1 Computational results based on feasibilty mapping and total cost function

Computational result of 4 instances in GA			
No. of feasible nurses	No. of waiting patients	Total wait time	Total costs
14	4	1.6	6.2
14	5	2.25	5
14	7	3.5	7.4
14	9	7.2	8.5

5 Conclusion

In this research paper the state of the art study of multiple nurse scheduling algorithms in healthcare has been presented and compared with a rank based rubric table. Finally the scope and the research gaps have been discussed in the direction to a proposed framework for remote nurse scheduling which has been designed here to improve the research challenges. A modified objective function for prioritized nurse assignment to a remote patient is formulated in the context of dynamic nurse scheduling framework. The objective function is finally implemented using genetic algorithm toolbox and results show that the best fitness value over a few generations is acceptable. Still there is a scope to implement penalty cost to consider the breakage of real soft constraints. This framework also needs to be implemented in an optimized manner and the flow of the proposed framework will be validated in an optimized way.

References

1. Burke, K.E., Li, J., Qu, R.: Pareto-based optimization for multi-objective nurse scheduling. In: Boros, E. (ed.) Annals of Operation Research, vol. 196, no. 1, pp. 91–109. Springer, US (2012). doi:10.1007/s10479-009-0590-8
2. Sarkar, P., Bhattacharya, U., Chaki, R., Sinha, D.: A priority based nurse scheduling algorithm for continuous remote patient monitoring. In: 4th World Conference on Applied Sciences, Engineering and Technology, pp. 046–053 (2015). ISBN 13: 978-81-930222-1-4
3. Sarkar, P., Sinha, D.: An approach to continuous pervasive care of remote patients based on priority based assignment of nurse. In: Saeed, K., Snášel, V. (eds.) 13th IFIP TC8 International Conference Computer Information Systems and Industrial Management, November 2014. CISIM 2014, LNCS 8838, pp. 327–338. Springer ISBN 978-3-662-45236-3, Online ISBN 978-3-662-45237-0. doi:10.1007/978-3-662-45237-0_31.Print
4. Métivier, J.-P., Boizumault, P., Loudni, S.: Solving nurse rostering problems using soft global constraints. In: Gent, I.P. (eds.) CP 2009, LNCS, vol. 5732, pp. 73–87. Springer, Berlin, Heidelberg (2009). doi:10.1007/978-3-642-04244-7_9
5. Moz, M., Pato, V.M.: A genetic algorithm approach to a nurse rerostering problem. Comput. Oper. Res. **34**, 667–691 (2007). doi:10.1016/j.cor.2005.03.019
6. Lim, J.G. et al.: Multi-objective nurse scheduling models with patient workload and nurse preferences. management. In: Bresciani, S. (ed.) vol. 2, no. 5, pp. 149–160. Scientific and Academic Publishing, p-ISSN: 2162-9374 e-ISSN: 2162-8416 (2012). doi:10.5923/j.mm.20120205.03

7. Trinkoff, Alison M., et al.: Nurses' work schedule characteristics, nurse staffing, and patient mortality. Nurs. Res. **60**(1), 1–8 (2011). doi:10.1097/NNR.0b013e3181fff15d. January/February

8. Aickelin, U., Dowsland, A.K.: An indirect genetic algorithm for a nurse scheduling problem. Comput. Oper. Res. Elsevier. **31**(5), 761–778 (2004). doi:10.1016/S0305-0548(03)00034-0

9. Dowsland, K.A., Thompson, J.M.: Solving a nurse scheduling problem with knapsacks, networks and tabu search. In: Crook, J., Archibald, T. (eds.) Journal of the Operational Research Society, vol. 51, no. 7, pp. 825–833. Springer (2000). doi:10.1057/palgrave.jors.2600970

10. Miller, H., Pierskalla, P.W., Rath J.: Nurse Scheduling using Mathematical Programming, Operations Research, vol. 24, no. 5 (1976). doi:10.1287/opre.24.5.857

11. Warner, M., Prawda, J.A.: Mathematical programming model for scheduling nursing personnel in a hospital. Management Science, vol. 19, no. 4, pp. 411–422 (1972). Application Series, Part 1. Published by: INFORMS

12. Howell, J.P.: Cyclical scheduling of nursing personnel. Hospitals Pubmed. **40**, 77–85 (1966)

13. Beasley, J.E., Cao, B.: A dynamic programming based algorithm for the crew scheduling problem. Comput. Oper. Res. Elsevier. **25**(7–8), 567–582 (1998). doi:10.1016/S0305-0548 (98)00019-7

14. Jaumard, B., Semet, F., Vovor, T.: A generalized linear programming model for nurse scheduling. Eur. J. Oper. Res. Elsevier. **107**(1), 1–18 (1998). doi:10.1016/S0377-2217(97) 00330-5

15. Satheesh kumar, B., Naresh kumar, S., Kumaraghuru, S.: Linear programming applied to nurses shifting problems for six consecutive days per week. Int. J. Curr. Res. **6**(03), 5862–5864. ISSN (Online): 2319-7064, March (2014)

16. Aickelin, U., White, P.: Building better nurse scheduling algorithms. In: Boros, E. (ed.) Annals of Operational Research, vol 128, no. 1, pp. 159–177. Springer (2004). p-SSN: 0254-5330, e-ISSN: 1572-9338

17. Fan, N., Mujahid, S., Zhang, J., Georgiev, P.: Nurse scheduling problem: an integer programming model with a practical application. In: Paradalos, P.M. et al. (ed.) Systems Analysis Tools for Better Health Care Delivery, Springer Optimization and Its Applications, vol. 74, pp. 65–98 (2012). doi:10.1007/978-1-4614-5094-8_5

18. Choy, M., Cheong L.F.M.: A Flexible Integer Programming framework for Nurse Scheduling. CoRR (2012). http://dblp.uni-trier.de/db/journals/corr/corr1210.html#abs-1210-3652. BibTeX key:journals/corr/abs-1210-3652

19. Maenhout, B., Vanhoucke, M.: Branching strategies in a branch-and-price approach for a multiple objective nurse scheduling problem. In: Burke, E. (eds.) Journal of Scheduling, vol. 13, no. 1,pp. 77–93. Springer, US (2010). doi:10.1007/s10951-009-0108-x

20. Darmoni, S.J., et al.: Horoplan: computer-assisted nurse scheduling using constraint-based programming. J. Soc. Health Care Pubmed. **5**, 41–54 (1995)

21. Cheng, B.M.W., Lee, J.H.M., Wu, J.C.K.: A nurse rostering system using constraint programming and redundant modeling. IEEE Trans. Inf. Technol. Biomed. **1**(1), 44–54 (1997). 10.1.1.48.2749

22. Chung Wong, G.Y., Chun, W.H.: Nurse rostering using constraint programming and meta-level reasoning. In: Hung, P.W.H., Hinde, C.J., Ali, M. (eds.) IEA/AIE 2003, LNAI, pp. 712–721 (2003)

23. Santos, D., Fernandes, P., Lopes, H.C., Oliveira, E.: A weighted constraint optimization approach to the nurse scheduling problem. In: IEEE 18th International Conference on Computational Science and Engineering, pp. 233–239 (2015). doi:10.1109/CSE.2015.46

24. Constantino, A.A., Landa-Silva, D, Melo, E.L., Xavier de Mendonc, D.F., Rizzato, D.B., Rom~ao, W.: A heuristic algorithm based on multi assignment procedures for nurse scheduling. In: Boros, E. (ed.) Journal, Annals of Operations Research, vol. 218, no. 1, pp. 165–183. Springer, US (2014). doi:10.1007/s10479-013-1357-9

25. Runarsson, T.P., Yao, X.: Stochastic ranking for constrained evolutionary optimization. IEEE Trans. Evol. Comput. **4**(3), 284–294 (2000). doi:10.1109/4235.873238

26. Bai, R., Burke, K.E., Kendall, G., Li, J., McCollum, B.: A hybrid evolutionary approach to the nurse rostering problem. IEEE Trans. Evol. Comput. **14**(4), 580–590 (2010). doi:10.1109/TEVC.2009.2033583. Aug
27. Brucker, P., Burke Edmund, K., Curtois, T., Qu, R., Berghe, V.G.: A shift sequence based approach for nurse scheduling and a new benchmark dataset. In: Laguna, M. (ed.) Journal of Heuristics August 2010, vol. 16, no. 4, pp. 559–573. Springer (2010). doi:10.1007/s10732-008-9099-6
28. Li, J., Aickelin, U.: Bayesian optimisation algorithm for nurse scheduling. In: Pelikan, M., Sastry, K., Cantu-Paz, E. (eds.) Scalable Optimization via Probabilistic Modeling: From Algorithms to Applications (Studies in Computational Intelligence), Chapter 17, pp. 315–332. Springer (2006)
29. Maenhout, B., Vanhoucke, M.: An electromagnetic meta-heuristic for the nurse scheduling problem. In: Laguna, M. (ed.) Journal of Heuristics, vol. 13, no. 4, pp 359–385. Springer (2007) doi:10.1007/s10732-007-9013-7
30. Jaszkiewicz, A.: A metaheuristic approach to multiple objective nurse scheduling. Found. Comput. Decis. Sci. **22**(3), 169–184 (1997)
31. Leksakul, K., Phetsawat, S.: Nurse scheduling using genetic algorithm. hindawi publishing corporation. Math. Probl. Eng. Article ID 246543, 16 (2014). http://dx.doi.org/10.1155/2014/246543
32. Moz, M., Pato, M.V.: A genetic algorithm approach to a nurse rerostering problem. Comput. Oper. Res. Elsevier. **34**(3), 667–691 (2007). doi:10.1016/j.cor.2005.03.019
33. Kim, S.-J., Ko, Y.-W., Uhmn, S., Kim, J.: A strategy to improve performance of genetic algorithm for nurse scheduling problem. Int. J. Soft. Eng. Appl. **8**(1), 53–62 (2014). 10.14257/Ijsela.14/8.1.05
34. Tsai, C., Li, A.H.S.: A two-stage modeling with genetic algorithms for the nurse scheduling problem. Expert Syst. Appl. **36**, 9506–9512 (2009). doi:10.1016/j.eswa.2008.11.049
35. Moscato, P., Cotta, C.: A modern introduction to memetic algorithms. Chapter 6. In: Gendreau, M., Potvin, J.-Y. (eds.) Handbook of Metaheuristics, International Series in Operations Research and Management Science, vol. 146, pp. 141–183. Springer, US (2010). doi:10.1007/978-1-4419-1665-56
36. Burke, K.E., Li, J., Qu, R.: A hybrid model of integer programming and variable neighbourhood search for highly-constrained nurse rostering problems. Eur. J. Oper. Res. Elsevier. **203**(2), 484–493 (2010). doi:10.1016/j.ejor.2009.07.036
37. Todorovic, N., Petrovic, S.: Bee colony optimization algorithm for nurse rostering. IEEE Trans. Syst. Man Cybern. Syst. **43**(2), 467–473 (2013). doi:10.1109/TSMCA.2012.2210404
38. Ghasemi, S., Sajadi, S.M., Vahdani, H.: Proposing a heuristic algorithm for the nurse scheduling in hospital emergency department (Case study: Shahid Beheshti Hospital). Int. J. Eng. Sci. **3**(9), 85–93 (2014). ISSN: 2306-6474
39. Ko, Y.W., Kim, D.H., Jeong, M., Jeon, W., Uhmn, S., Kim, J.: An efficient method for nurse scheduling problem using simulated annealing. In: The 5th International Conference on Advanced Science and Technology, AST 2013, vol. 20, pp. 82–85. ASTL (2013)
40. Ko, Y.W., Kim, D.H., Jeong, M., Jeon, W., Uhmn, S., Kim J.: An improvement technique for simulated annealing and its application to nurse scheduling problem. Int. J. Soft. Eng. Appl. **7**(4), 269–277 (2013)
41. Stepanov, A.: On the Borel-Cantelli Lemma. Department of Mathematics, Izmir University of Economics, Turkey. AMS 2000 Subject Classification: 60G70, 62G30 (2000)

A Divide-and-Conquer Algorithm for All Spanning Tree Generation

Maumita Chakraborty, Ranjan Mehera and Rajat Kumar Pal

Abstract This paper claims to propose a unique solution to the problem of all possible spanning tree enumeration for a simple, symmetric, and connected graph. It is based on the algorithmic paradigm named divide-and-conquer. Our algorithm proposes to perform no duplicate tree comparison and a minimum number of circuit testing, consuming reasonable time and space.

Keywords Circuit testing · Connected graph · Divide-and-conquer · Duplicate tree comparison · Simple graph · Spanning tree · Symmetric graph

1 Introduction

Divide-and-conquer is a well-known algorithmic approach for solving different problems. It has three phases namely, divide, conquer, and combine [1]. In this paper, we have used this new divide-and-conquer approach in generating all possible spanning trees of a simple, symmetric, and connected graph. All spanning tree generation has been an area, well-explored and well-known for its wide applications in the fields of computer science, chemistry, medical science, biology and many others. The algorithms developed for spanning tree enumeration are mainly targeted towards generating all trees in optimum time and space. Moreover, checking for the

M. Chakraborty (✉)
Department of Information Technology, Institute of Engineering
and Management, Y-12, Block-EP, Sector-V, Salt Lake, Kolkata, India
e-mail: maumita.chakraborty@gmail.com

R. Mehera
Subex, Inc., 12303 Airport Way, Suite 390, Broomfield, CO 80021, USA
e-mail: ranjan.mehera@gmail.com

R.K. Pal
Department of Computer Science and Engineering, University of Calcutta,
JD-2, Sector III, Salt Lake, Kolkata, India
e-mail: pal.rajatk@gmail.com

© Springer Nature Singapore Pte Ltd. 2017
R. Chaki et al. (eds.), *Advanced Computing and Systems for Security*,
Advances in Intelligent Systems and Computing 567,
DOI 10.1007/978-981-10-3409-1_2

duplicate tree and non-tree sequences are two major issues in this area of research. In our algorithm, we claim to eradicate the duplicate tree issue completely and also reduce the number of non-tree sequence generations.

2 Literature Survey

Starting from the 60s till date, many algorithms have been devised for computation of all possible trees of a simple, symmetric, and connected graph. Based on the methods being used, three major classifications have been made in this regard, namely: tree testing method [2–11], elementary tree transformation method [12–19], and trees by successive reduction method [20–22].

All the algorithms have some unique features (as well as novelties and limitations) of their own, which may differ even when they are under the same head of classification. In general, *tree testing method* is mainly responsible for generating all possible sequences of the desired length, among which the tree sequences are accepted, and the non-trees are discarded. On the other hand, the second category, namely *elementary tree transformation* starts with an initial BFS or DFS tree and then repeatedly generates one tree from the other by replacement of one or more edges, whose selection criteria is different for different algorithms. Lastly, the *successive reduction method* reduces the graph to trivial subgraphs gradually. Trees of the original graph are then obtained from the trees of the trivial subgraphs. Almost all the algorithms till date can somehow be categorized under any one of the above, as a result of which there is always a scope for devising some new approach in this problem domain.

In the subsequent sections, we discuss the functioning of some of the earlier existing methods of generating all spanning trees in brief, then the development of our algorithm, *DCC_Trees*, with proper examples and elaborations.

3 Existing Techniques for All Spanning Tree Generation

In this section, we present very briefly the underlying working principle of three algorithms which fall under the three existing techniques of all spanning tree generation, as mentioned above.

3.1 Trees by Test and Select Method

As mentioned in the earlier section, this method generates all possible sequences of edges; some of them are tree sequences while others are not. J.P. Char had adopted this method of tree generation in his algorithm. Char defines a $(n - 1)$-digit sequence $\lambda =$ (DIGIT(1), DIGIT(2), ..., DIGIT($n - 1$)) such that DIGIT(i), $1 \leq i \leq n - 1$, is a vertex adjacent to vertex i in G. He has also given a *Tree Compatibility Checking* for the sequences as:

The sequence (DIGIT(1), DIGIT(2), ..., DIGIT($n - 1$)) represents a spanning tree of graph G if and only if for each $j \leq n - 1$, there exists in G a sequence of edges with (j, DIGIT(j)) as the starting edge, which leads to a vertex $k > j$ [3].

Here we present Char's algorithm [3] at a glance.

Any graph G is represented by the adjacency lists of its vertices. SUCC(DIGIT(i)) is the entry next to DIGIT(i) in the adjacency list of vertex i.

1. Begin
2. Find the initial spanning tree and obtain the initial tree sequence $\lambda =$ (REF(l), REF(2), ..., REF($n - 1$));
3. Renumber the vertices of the graph using the initial spanning tree;
4. Initialize DIGIT(i): = REF(i), $1 \leq i \leq n - 1$;
5. Output the initial spanning tree;
6. k: = $n - 1$;
7. While $k \neq 0$ do begin
8. If SUCC(DIGIT(k)) $\neq$ nil then
9. Begin
10. DIGIT(k): = SUCC(DIGIT(k));
11. if DIGIT(i), $1 \leq i \leq n - 1$, is a tree sequence then
12. Begin
13. Output the tree sequence;
14. k: = $n - 1$;
15. End;
16. End;
17. Else begin
18. DIGIT(k): = REF(k);
19. k: = $k - 1$;
20. End;
21. End;

For the example graph G' shown in Fig. 1, we can start with an initial tree ((v_1, v_2), (v_2, v_4), (v_4, v_3)) obtained by performing BFS on G'. The vertices v_1, v_2, v_4, and v_3 are renumbered to v_1, v_2, v_3, and v_4, respectively. The adjacency lists of the graph show that there are no more adjacent vertices from v_4 or v_2. Hence, the next tree sequence will be ((v_1, v_4), (v_4, v_2), (v_3, v_1)) and then ((v_1, v_4), (v_2, v_1), (v_4, v_3)), and so on.

Fig. 1 An example graph G'

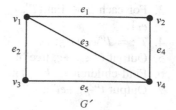

G'

3.2 Elementary Tree Transformation Method

It has been mentioned earlier that this technique mainly operates by replacing one edge from an existing tree by another suitable edge to give rise to another tree. One such algorithm by Shioura and Tamura [18], following the above working principle, assumes that any spanning tree T^0 be the progenitor of all spanning trees. It then outputs all spanning trees by reversely scanning all children of any spanning tree. They have also defined a useful child-parent relation.

Let G (consisting of V vertices and E edges) be an undirected connected graph consisting of a vertex set $\{v_1, v_2, ..., v_V\}$ and an edge set $\{e_1, e_2, ..., e_E\}$. A spanning tree of G is represented as its edge-set of size $(V - 1)$. For any spanning tree T and any edge $f \in T$, the subgraph induced by the edge-set $T\backslash f$ has exactly two components. The set of edges connecting these components is called the fundamental cut associated with T and f, written as $C^*(T\backslash f)$. For any edge $f \in T$ and any arbitrary edge $g \in C^*(T\backslash f)$, $T\backslash f \cup g$ is also a spanning tree. For any edge $g \notin T$, the edge-induced subgraph of G by $T \cup g$ has a unique circuit, the fundamental circuit associated with T and g. The set of edges of the circuit is denoted by $C(T \cup g)$. For any $g \notin T$ and any $f \in C(T \cup g)$, $T \cup g\backslash f$ is a spanning tree. Relative to a spanning tree T of G, if the unique path in T from vertex v to the root v_1 contains a vertex u, then u is called an ancestor of v and v is the descendant of u. Similar is the case for edges. A depth-first spanning tree has been defined as a spanning tree such that for each edge of G, its one incidence vertex is the ancestor of the other. The algorithm at a glance, as given by the authors [18] is as follows.

Procedure All-Spanning-Trees (G)

Input: a graph G with a vertex-set $\{v_1, v_2, ..., v_V\}$ and an edge-set $\{e_1, e_2, ..., e_E\}$;

1. Begin
2. Using depth-first search, do
    - find a depth-first spanning tree T^0 of G;
    - sort vertices and edges to satisfy the ancestor-descendant relationship;
3. Output T^0 {"$e_1, e_2, ..., e_{V-1}$, tree"};
4. Find-children $(T^0, V - 1)$;
5. End;

Procedure Find-Children (T^p, k)

Input: a spanning tree T^p and an integer k with $e_k < \text{Min}(T^0\backslash T^p)$;

1. Begin
2. If $k \leq 0$ then return;
3. For each $g \in \text{Entr}(T^p, e_k)$ do begin {output all children of T^p not containing e_k};
4. $T^c := T^p\backslash e_k \cup g$;
5. Output ("$-e_k, +g$, tree");
6. Find-children$(T^c, k - 1)$;
7. Output ("$-g, +e_k$");

8. End;
9. Find-children(T^p, $k - 1$);{find the children of T^p containing e_k}
10. End;

Following the above algorithm, $T^0 = (e_1, e_4, e_5)$ can be an initial DFS tree for the graph G' (Fig. 1). The edges e_1, e_4, and e_5 are to be replaced by other suitable edges of the graph one by one to give rise to its child trees. For example, e_1 can be replaced by e_2 and e_3 to give rise to the trees T^{01} (e_4, e_5, e_2) and T^{02} (e_4, e_5, e_3). Similarly, replacement of e_4 and e_5 will result in some more second generation trees. Once again, the second generation trees like T^{01}, T^{02}, etc. can also give rise to some other trees by replacement of their edges by some other suitable edges of the graph G'. This process of tree generation stops as no more child trees are generated.

3.3 Successive Reduction Method

This tree generation method operates on the principle of dividing a large graph into smaller subgraphs continuously till the reduced graphs are trivial like edges. Trees of the original graph are then obtained from the trees of the trivial subgraphs. One of the algorithms which rely on this technique is by Winter [22]. Here, a new enumeration algorithm on spanning trees based on the idea of contraction and removal of edges of the graph is presented. First, it constructs all spanning trees containing some selected edge e_1, then all spanning trees containing another edge e_2, but not e_1, and so on.

Figure 2 shows the computation flow tree for generating all spanning trees of the graph G' (of Fig. 1) using Winter's algorithm. The labels on the edges of the

Fig. 2 Computation flow tree for Winter's tree generation

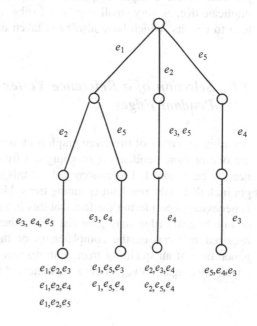

computation tree are for those edges of G' along which contraction takes place. For example, when G' is contracted along e_1, it results into two branches (left one indicates further contraction along e_2 and right one indicates further contraction along e_5). That is why, the trees generated from the leftmost branch always include e_1 and e_2, while the next branch always includes e_1 and e_5, but never includes e_2. Similarly, contractions on G' can start from two more edges, namely e_2 and e_5 resulting into two different sets of trees which never include e_1 as one of their edges. The trees generated following a particular sequence of contractions are mentioned at the bottom of each path in the computation flow tree of graph G', shown in Fig. 2.

4 The New Algorithm (*DCC_Trees*)

Our objective is to use divide-and-conquer, a new paradigm towards the generation of all possible spanning trees of a graph [1]. A typical divide-and-conquer algorithm solves a problem using following three steps:

Divide: Break the given problem into subproblems of the same type.
Conquer: Recursively solve these subproblems.
Combine: Appropriately combine the results of the subproblems.

In our algorithm, *DCC_Trees*, we have used the approach mentioned above to divide a graph into partitions, based on certain criteria/measures, and then merged them in various possible ways to generate different spanning trees. We have defined two types of partitions: primary and secondary, depending on the necessity of their presence in the computed spanning trees. Our algorithm claims to generate no duplicate tree. A very small number of subgraphs generated by the algorithm may lead to circuits, which have also been taken care of subsequently.

4.1 Selection of a Reference Vertex and Elimination of Its Pendant Edges

An arbitrary vertex of the given graph is chosen as the reference vertex, v_{ref}. If there are one or more pendant edges going out from v_{ref} in the graph, then those edges need to be removed. The concept behind this removal is that pendant edges always get included in the resultant spanning trees. Moreover, the removal of pendant edge is necessary, considering the fact that this is a recursive procedure because removal of one pendant edge may give rise to another one and so forth. This step is also required in terms of the completeness of the proposed algorithm. Hence, after generation of all spanning trees with respect to v_{ref}, we need to add the pendant edges once again to v_{ref} to get v_{ref} connected to those vertices which are the other

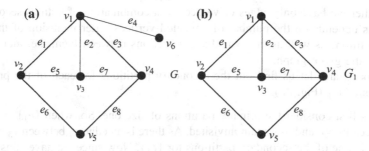

Fig. 3 **a** An example graph, G **b** Graph G_1 derived from G by removing pendant edge e_4

end vertices of those pendant edges. In this regard, let us consider an example graph G in Fig. 3a. Let us choose any vertex v_1 as v_{ref}. From v_1, there is a pendant edge e_4 going out to v_6, which is removed for the time being. As a result, the graph formed (G_1) is shown in Fig. 3b. We now compute all possible trees of the graph G_1.

4.2 Divide: Decomposition of the Graph

In this phase of the algorithm, we divide the graph into partitions, based on certain criteria/measures. The partitions have been classified into two types: primary and secondary. The formation of primary and secondary partitions is being described in the following two subsections.

Formation of primary partitions. All possible combinations of the edges coming out from the reference vertex, v_{ref}, taken one to all at a time, form the primary partitions in our algorithm. A primary partition is a necessary or mandatory component of the resultant spanning trees formed from the given graph, as it will ensure the connectivity of v_{ref} with the rest of the graph. If there are m edges coming out of v_{ref}, then the number of primary partitions from v_{ref} is the sum $^mC_1 + ^mC_2 + \cdots + ^mC_m$.

For the example graph, G_1 considered in Fig. 3b, the three edges coming out of v_1 (v_{ref} for G_1) are e_1, e_2, and e_3. Thus, the seven ($^3C_1 + ^3C_2 + ^3C_3$) primary partitions for G_1 are $\{e_1\}$, $\{e_2\}$, $\{e_3\}$, $\{e_1, e_2\}$, $\{e_1, e_3\}$, $\{e_2, e_3\}$, and $\{e_1, e_2, e_3\}$.

Formation of secondary partitions. Once a primary partition is selected, one or more vertices, other than the v_{ref}, also get included or visited. Some other unvisited vertices may still be there in the graph. Each edge between a pair of distinct unvisited vertices is considered to be a separate secondary partition. Not only edges, but single left-out vertices can also form separate secondary partitions. After selecting any one primary partition, we find out the secondary partitions, as required. These are the conditional partitions which are subject to change with a change in a primary partition.

Whether we have only edges or vertices or a combination of both as secondary partitions depends on the number of unvisited vertices (after inclusion of the primary partition) as well as the available connections between them, i.e. their adjacency in the given graph.

In this regard, let us find out the secondary partitions for each of the primary partitions of G_1 (Fig. 3b).

- Let us first consider the primary partitions of size one. So, when $\{e_1\}$ is taken, vertices v_3, v_4, and v_5 remain unvisited. As there is an edge e_7 between v_3 and v_4, so e_7 is one of the secondary partitions for $\{e_1\}$. Now since we have considered edge e_7, which implies we have already visited vertices v_3 and v_4 and hence edge e_8 can't be considered as a secondary partition, since one end of e_8 is already visited (vertex v_4). So now there is a left-out vertex v_5 which has not been included. As a result, v_5 becomes another secondary component. Similarly, for each of the next two primary partitions, $\{e_2\}$ and $\{e_3\}$, we get one edge component and one vertex component as the secondary partitions. For $\{e_2\}$, they are either e_6 and v_4, or e_8 and v_2 respectively. And for $\{e_3\}$, the secondary partitions are either e_5 and v_5, or e_6 and v_3, e_8 and v_2 respectively.
- Then, we consider the primary partitions of size two. So, for $\{e_1, e_2\}$ and $\{e_2, e_3\}$, e_8 and e_6 form the secondary partitions, respectively. However, for $\{e_1, e_3\}$, we do not get any edge component, as the left-over vertices v_3 and v_5 are not connected. Hence, v_3 and v_5, in this case, form two separate secondary partitions.
- Lastly, for the primary partition $\{e_1, e_2, e_3\}$ of size three, vertex v_5 is the only secondary partition.

4.3 Conquer: Searching for Connectors

This section is targeted towards finding out what are the different ways in which the partitions can be joined to give rise to all possible and different spanning trees.

Once we have all the partitions available from the divide phase of *DCC_Trees*, we search for the connectors which are supposed to join them. For a particular primary partition, the secondary partitions are gradually joined in the reverse order in which they were formed by a set of connectors named as CS_1, CS_2, ..., and so on. A pair of partitions along with the connectors from the corresponding connector set gives rise to the subgraphs which gradually lead to the formation of different spanning trees of the graph. This process is continued till we find out the connectors that join the above-formed subgraph with the corresponding primary partition, i.e. those edges which join the primary partition with the rest of the vertices in the given graph. In case any secondary partition, s_1, does not find a connection with any other secondary partition or s_1 is the singleton secondary partition, we directly join s_1 with the corresponding primary partition with some bridges named as main connectors. *MC* is such a set of main connectors.

Let us now find out the set of connectors for the example graph G_1 in Fig. 3b.

- For partition $\{e_1\}$, CS_1 (joining e_7 with v_5) = e_8, and MC_1 (joining e_1 with rest) = (e_5, e_6). Similarly, for $\{e_2\}$, CS_1 (joining e_6 with v_4) = e_8, and MC_1 (joining e_2 with rest) = (e_5, e_7). Also, for $\{e_3\}$, CS_1 (joining e_5 with v_5) = e_6, and MC_1 (joining e_3 with rest) = (e_7, e_8).
- Then, we consider the next set of partitions of size two. For each of $\{e_1, e_2\}$ and $\{e_2, e_3\}$, there is only one secondary partition; hence, no CS's. So, MC_1 (for $\{e_1, e_2\}$) = (e_6, e_7) and MC_1 (for $\{e_2, e_3\}$) = (e_5, e_8). However, for the partition $\{e_1, e_3\}$, there are two secondary partitions, each of which has to be joined separately with $\{e_1, e_3\}$. Hence, MC_1 (joining $\{e_1, e_3\}$ with v_3) = (e_5, e_7) and MC_2 (joining $\{e_1, e_3\}$ with v_5) = (e_6, e_8).
- For the last partition $\{e_1, e_2, e_3\}$, the only MC_1 (joining $\{e_1, e_2, e_3\}$ with v_5) = (e_6, e_8).

Thus, in the conquer phase, for each primary partition, we find out all possible sets of connectors for all pairs of partitions until we include all the vertices of the graph. Now, we move to combine phase of our algorithm, where we decide how to take different combinations of connectors for each set of partitions to form different tree sequences.

4.4 Combine: Forming the Trees

In this section, we form the required trees of the given graph in two phases.

Tree generation: Phase 1. From the earlier two sections, we know that, for each primary partition, p_i, there can be a set of secondary partitions, S_i, where $|S_i| \geq 0$. To join each such secondary partition of S_i with one other, there are sets of connectors, CS_i, and to join them with the primary partitions, we have a set of main connectors, MC_i, where $|MC_i| \geq 0$. We take two partitions and one connector from the corresponding CS, at a time, to form a subgraph, which is again joined with another partition taking one connector from the next set, CS. This continues till the primary partition is joined with this gradually growing subgraph using one main connector from set MC. The largest subgraph thus formed is a spanning tree of the given graph. Again, the same process is repeated with new combinations of connectors at each step, giving rise to a new tree. Taking different combinations of connectors from CS and MC, we thus get different trees of the graph. The whole procedure is repeated for each p_i, thus, yielding more trees in each case. The algorithm claims to generate no circuit or duplicate tree in this phase.

Following the above procedure of Phase 1 tree generation, we get the following trees; here the trees are formed by exactly $n - 1$ edges of a given graph enclosed within parentheses:

- The partition $\{e_1\}$ of G_1 yields the following trees: (e_1, e_7, e_8, e_5) and (e_1, e_7, e_8, e_6), where e_1 is the primary partition, e_7 is secondary partition, e_8 is the connector joining e_7 with v_5, e_5 and e_6 are two main connectors joining $\{e_1\}$ with the rest of the subgraph. Similarly, $\{e_2\}$ generates the trees (e_2, e_6, e_8, e_5) and (e_2, e_6, e_8, e_6), while $\{e_3\}$ generates (e_3, e_5, e_6, e_7) and (e_3, e_5, e_6, e_8).

- Trees from partition $\{e_1, e_2\}$: (e_1, e_2, e_8, e_6) and (e_1, e_2, e_8, e_7). Trees from $\{e_1, e_3\}$: (e_1, e_3, e_5, e_6), (e_1, e_3, e_5, e_8), (e_1, e_3, e_7, e_6) and (e_1, e_3, e_7, e_8). Also, trees from $\{e_2, e_3\}$: (e_2, e_3, e_6, e_5) and (e_2, e_3, e_6, e_8).
- Lastly, partition $\{e_1, e_2, e_3\}$ generates the trees (e_1, e_2, e_3, e_6) and (e_1, e_2, e_3, e_8).

Tree generation: Phase 2. This is the next phase in tree generation where we take more than one connector from each set, whether *CS* or *MC*. If there are *r* connectors in a connector set, then we take combinations of two or more (up to *r*) connectors at a time. The salient features of this phase are:

1. All the partitions are not included here. Only the primary partition and the corresponding connector combination from *CS* are taken. These now become the mandatory component for this particular case. With the remaining unvisited vertices, we once again form secondary partitions. The mandatory component and the secondary partitions can thus be combined in all possible ways, as described in Phase 1 of tree generation. This is a recursive process where the new combination of two or more connectors is added to the primary partition to form the new mandatory component. Once again secondary partitions are found out and once again they are joined by connectors like before.
2. If there are more than two connectors in any connector set, *CS* or *MC*, we take combinations of two or more such connectors. Now this connector combination is joined with the remaining secondary partition(s) by another set of connectors which should not include any member of the previous connector set, whose members were already considered.
3. While taking combinations of connectors from any connector set, we do not take those combinations which when taken along with the primary partition do not exceed the size of a tree of the given graph.
4. When the main connectors are combined, the primary partition and the main connectors form the only mandatory component at that time, and the same process of finding out secondary components, joining them with the mandatory component in all possible ways is repeated. However, this is the only phase where there is a chance of circuit formation.
5. While combining the main connectors of *MC*, if we find at least two connectors from two different end points of the primary partition, going out to a common vertex, a circuit is formed. Hence, that particular combination of connectors does not yield any tree and thus needs to be discarded.

Theorem 1 *The algorithm never generates duplicate trees.*

Proof According to the algorithm, Phase 1 takes different connectors from different *CS*'s and *MC*'s at a time. Thus, no question of duplication arises. In Phase 2 of tree generation, we take two or more combinations of connectors from the same *CS* or *MC* along with the primary partition to generate more and more trees. This process may get repeated based on new sets of connectors found at each step. For a particular primary partition, no connector combination from any connector set is allowed to have a member of its own set which joins the remaining secondary component(s). But, if the primary partition changes, and if some connector combination repeats, it does not generate any duplicate tree because of inclusion or exclusion of different edges in the primary partition.   □

Lemma 1 *Circuits are formed only for primary partitions of size k > 1.*

Proof A primary partition of size $k > 1$ has k number of branches going out from v_{ref}, and hence, k end points. There can be k or even more number of connectors, from k such end points going out to a common unvisited vertex of the graph. Even if there are at least two different connectors from two end points going out to a common vertex, there is a circuit. Conversely, if $k = 1$, i.e. there is a single end point of the primary partition, no question of circuit formation arises.   □

Lemma 2 *A primary partition of size k > 1, does not necessarily mean that there is a circuit.*

Proof From Lemma 1, it is proved that circuits are formed for only $k > 1$, but the reverse is not always true. If there are k end points of the primary partition and if two or more connectors come out of the same end point but goes to different unvisited vertices of the graph, no circuit is formed. But, if at least two such connectors coming out of different end points go to the same unvisited vertex, a circuit is formed. Hence, Lemma 2 is proved.   □

Corollary 1 *Joining a primary partition of size k > 1 with a vertex component always yields circuit, while joining it with an edge component may or may not yield circuit.*

Proof From Lemmas 1 and 2, we have seen that primary partitions of size $k > 1$ may or may not generate circuits. If the primary partition is being joined with a vertex component, connector combinations from two or more (up to k) end points always go to the vertex, forming a circuit. On the other hand, if the primary partition is being joined with an edge component, then the connectors from the different endpoints may go to the same end-vertex of the edge, forming circuits or to different end vertices of the edge which does not form a circuit.   □

Let us explain the Phase 2 tree generation for the graph G_1 of Fig. 3b.

- For partition $\{e_1\}$, only one connector is there in *CS*. Main connector combination gives (e_5, e_6). Only v_4 is left unvisited, which can be joined by both e_7 and e_8. Thus, trees formed here are: (e_1, e_5, e_6, e_7) and (e_1, e_5, e_6, e_8). Similarly, for partition $\{e_2\}$, trees formed are: (e_2, e_5, e_7, e_6) and (e_2, e_5, e_7, e_8), and for partition $\{e_3\}$, trees are: (e_3, e_7, e_8, e_5) and (e_3, e_7, e_8, e_6).
- Next, for $\{e_1, e_2\}$, no *CS* is there, as there is a single secondary partition. Main connector combination gives (e_6, e_7). Hence, tree formed: (e_1, e_2, e_6, e_7). Similarly, for $\{e_2, e_3\}$, the tree formed is: (e_2, e_3, e_5, e_8). On the contrary, for partition $\{e_1, e_3\}$, there are two *MC*'s, MC_1 joining v_3 with $\{e_1, e_3\}$ and MC_2

joining v_5 with $\{e_1, e_3\}$. Since, both v_3 and v_5 are vertex-components, so combination of connectors in MC_1 as well as MC_2 generates circuit, which are thus not considered.

- Lastly, for $\{e_1, e_2, e_3\}$, only the vertex v_5 remains unvisited, which is once again a vertex component. Hence, the connectors in MC, namely, e_6 and e_8, are not combined, else circuit is formed.

4.5 Rejoining the Pendant Edges

In the last section, we rejoin the pendant edges from v_{ref} to all the trees obtained in the previous sections, to get the spanning trees of the original graph. After generating the trees for the graph G_1 (in Fig. 3b) in the earlier sections, we join the pendant edge (e_4) from $v_{ref}(v_1)$ with them. These give us the following trees for the original graph G (Fig. 3a): $(e_1, e_7, e_8, e_5, e_4), (e_1, e_7, e_8, e_6, e_4), (e_2, e_6, e_8, e_5, e_4),$ $(e_2, e_6, e_8, e_6, e_4), (e_3, e_5, e_6, e_7, e_4), (e_3, e_5, e_6, e_8, e_4), (e_1, e_2, e_8, e_6, e_4), (e_1, e_2, e_8,$ $e_7, e_4), (e_1, e_3, e_5, e_6, e_4), (e_1, e_3, e_5, e_8, e_4), (e_1, e_3, e_7, e_6, e_4), (e_1, e_3, e_7, e_8, e_4), (e_2,$ $e_3, e_6, e_5, e_4), (e_2, e_3, e_6, e_8, e_4), (e_1, e_2, e_3, e_6, e_4), (e_1, e_2, e_3, e_8, e_4), (e_1, e_5, e_6, e_7,$ $e_4), (e_1, e_5, e_6, e_8, e_4), (e_2, e_5, e_7, e_6, e_4), (e_2, e_5, e_7, e_8, e_4), (e_3, e_7, e_8, e_5, e_4), (e_3, e_7,$ $e_8, e_6, e_4), (e_1, e_2, e_6, e_7, e_4),$ and $(e_2, e_3, e_5, e_8, e_4)$.

4.6 Algorithm DCC_Trees

1. **Begin**
2. Take an input graph $G = (V, E)$.
3. Take any vertex from V as v_{ref}.
4. Remove pendant edges from v_{ref}, if any, to form graph G_1.
5. Take all possible combinations of all edges from v_{ref}, including one to all edges at a time, forming primary partitions (say, $P_1, P_2, ..., P_k$).
6. $\forall P_i, 1 \leq i \leq k$ **begin**
7. Form secondary partitions (say, $S_1, S_2, ..., S_w$) with either unvisited vertices or edges joining unvisited vertices.
8. For $j = w$ to 1 **begin**
9. Consider a pair of secondary partitions (S_j, S_{j-1}).
10. Find out the set of connector(s) (if exist(s)), CS, joining (S_j, S_{j-1}).
11. If $CS = \phi$, go to Step 15.
12. For each such CS **begin**
13. Join it with the pair (S_j, S_{j-1}) forming subgraph SB.
14. $S_j \leftarrow$ Subgraphs formed in Step 10.
15. **End**
16. $j \leftarrow j - 1$.

17. **End**
18. Unjoined secondary partitions (if any) joined to P_i to form sequences in the tree set SP of Phase 1.
19. Take combinations of two or more connectors from each connector set, CS or MC, along with primary partition, and then form secondary partitions, if required, to generate trees of Phase 2.
20. **End**
21. Add pendant edges of v_{ref} to trees generated in Steps 7 through 9 to generate trees of G.
22. **End**

Theorem 2 *DCC_Trees generates all possible spanning trees of a simple, connected graph.*

Proof DCC_Trees generates primary partitions from a reference vertex, v_{ref}. Primary partitions consider all combinations of edges incident on v_{ref}, ensuring inclusion of v_{ref} in all the sequences generated as well as considering all the sequences exhaustively with a different set of edges from v_{ref}. For each such primary partition, all the edges encompassing adjacent unvisited vertices as well as left-out unvisited vertices, named as secondary partitions, are being considered. Moreover, all the connectors joining the different secondary partitions along with the corresponding primary partition is also considered in this algorithm. If there is no connector between a pair of secondary partitions, then the next secondary partition, in sequence, is considered. Even if there is a secondary partition which has no connection with any other secondary partition being formed, it is directly joined with the primary partition with a set of main connectors. The algorithm thus carries out exhaustive divide and conquer phases, where all components and all possible ways of joining them are considered, ensuring generation of all possible spanning trees of the given graph.

4.7 Data Structures and Complexity Issues

The algorithm stores the given graph, G, in an adjacency matrix, whose storage requirement is $O(n + m)$, where n is the number of vertices and m is the number of edges of the graph. The primary partitions are being stored in a linked list, one at a time, and hence requires $O(n)$ space only. Searching for v_{ref} in the given graph and removal of its pendant edges can take at most $O(n)$ time. In the division phase, for a particular primary partition, forming the other partitions can take at most $O(n)$ time. There can be at most $\lceil n/2 \rceil$ components having maximum two end points. Thus, searching for the connectors takes at most $O(n)$ time. Combining such connectors and partitions also require $O(n)$ time. Now, there can be at most $O(2^n)$ primary partitions, when the maximum degree of a vertex is $n - 1$, i.e. it is a complete graph. The maximum time taken to generate trees for any one of them is $O(n)$. Hence, the worst case time complexity of the algorithm comes out to be $O(n.2^n)$.

4.8 Salient Features of DCC_Trees

In this section, we like to highlight on some of the key features of the algorithm as follows.

- The algorithm is based on a unique approach never attempted before.
- It guarantees no duplicate tree generation.
- It can generate the spanning trees with a particular edge(s) included, i.e., if we want to find out those spanning trees of a given graph with a particular edge or a group of edges from a particular vertex being always included, then our algorithm can do so efficiently.
- The consistency of the proposed algorithm remains valid irrespective of the selection of starting vertex.
- It is also suitable for parallel processing. Once the primary partitions are ready, trees generated from any one partition are independent of those from another partition, and hence, can be processed in parallel.

5 Experimental Results

Even if the computation time required for computing all spanning trees is exponential, we execute the algorithm devised in this paper, DCC_Trees, for graphs whose order as well as size is bounded by some constant, and we can generate all trees for each of such assumed instances in feasible/reasonable amount of time (and space).

In this section, we have incorporated the implementation results of our algorithm for random graph instances having $|V| = 10$ to 22, $|V|$ being the total number of vertices of the graph instances. The implementation has been carried out on an Intel Core i3 quad-core processor running at 2.4 GHz, with 6 GB RAM capacity. A few standard algorithms for generating all possible spanning trees, given by Shioura and Tamura [18], Matsui [15], Mayeda and Seshu [17], Hakimi [13], Char [3], and Winter [22] have also been implemented in the same environment and on same graph instances. Table 1 gives the order and size of the graph instances considered, number of trees generated, and the CPU time taken by all the algorithms to run each instance of the graph.

In Table 1, $I_i(x, y)$ is the i-th instance of a graph of order x (i.e. number of vertices) and size y (i.e. number of edges). The time taken by each of the algorithms is shown in an mm-ss format, where mm and ss stand for minutes and seconds required to execute the algorithms on the specific instances, respectively. A few observations from Table 1:

Table 1 Experimental results of computing all spanning trees for random graph instances, having order from 10 to 22

Instances with vertex# and edge#	Number of trees generated	Shioura and Tamura [18]	Matsui [15]	Mayeda and Seshu [17]	Hakimi [13]	Char [3]	Winter [22]	DCC_Trees
		(mm-ss)	(mm-ss)	(mm-ss)	(mm-ss)	(mm-ss)	(mm-ss)	(mm-ss)
I_1 (10, 18)	6210	00-00	00-00	00-00	00-00	00-01	00-00	00-00
I_2 (10, 15)	636	00-00	00-00	00-00	00-00	00-00	00-00	00-00
I_3 (10, 14)	364	00-00	00-00	00-00	00-00	00-00	00-00	00-00
I_1 (15, 23)	6054	00-00	00-00	00-00	00-00	00-01	00-00	00-00
I_2 (15, 21)	1320	00-00	00-00	00-00	00-00	00-00	00-00	00-00
I_3 (15, 21)	2858	00-00	00-00	00-01	00-00	00-00	00-00	00-00
I_1 (20, 35)	13100220	09-35	02-20	25-30	08-24	09-41	00-38	00-43
I_2 (20, 28)	32854	00-03	00-01	00-05	00-01	00-05	00-00	00-01
I_3 (20, 31)	248120	00-17	00-03	00-33	00-10	05-43	00-02	00-05
I_1 (22, 32)	616642	01-17	00-13	02-28	01-14	12-04	00-07	00-23

- In the above table, we have shown three instances for each of 10, 15 and 20 vertices and one instance of 22 vertex graph.
- The density of the graphs (actual number of edges in the graph/maximum number of edges possible with given number of vertices) ranges from 0.1 to 0.4 approximately.
- A number of trees generated depend not only on the order and size of the graph but also on the arrangement of edges in the graph.
- Time taken to execute the algorithms depends not only on order and size of the graph but also on the number of trees generated.
- It is observed that the algorithm by Winter [22] gives best results with respect to CPU time compared to all other algorithms considered here. Winter proposes to generate several different tree sequences simultaneously.
- The last column in the table shows the execution time of our algorithm, DCC_Trees, which is found to be better than most of the others. In our algorithm, Phase 1 generates several tree sequences simultaneously, thus consuming much less time than that of Phase 2.

6 Applications

Many problems in various fields of science and engineering need to be formulated regarding graphs. Many of them involve various applications of spanning trees like computation of minimum spanning tree or generation of all possible spanning trees from a given graph.

Some of the application areas of spanning trees are as follows:

- *Network Design*: Designing different networks such as phone, electrical, hydraulic, TV cable, computer, road, air traffic, railway, electronic circuits, etc.
- *Approximation Algorithms for NP-hard Problems*: Solving problems like the travelling salesman problem (TSP) having several applications in planning, logistics, and the manufacture of microchips, as well as in DNA sequencing.
- *Cluster Analysis*: Required for routing in mobile ad-hoc networks, identifying patterns in gene expression, document characterization for web search, medical image processing.
- *Image Processing*: For extraction of networks of narrow curvilinear features such as road and river networks from remotely sensed images.
- *Astronomy and Space Sciences*: To compare the aggregation of bright galaxies with faint ones.
- *Biology*: To carry out research in the quantitative description of cell structures in light microscopic images.
- *Molecular Biology*: Frequently used in molecular epidemiology research to estimate relationships among individual strains or isolates.
- *Chemistry*: Used in chemical research for determination of the geometry and dynamics of compact polymers.
- *Archaeology*: For identifying proximity analysis.
- *Bioinformatics*: For micro-array expression of data.
- *Geography*: Used in efficient methods for regionalization of socio-economic units in maps.
- *Binary Spanning Trees*: It is a rooted structure with a parent having 0, 1, or 2 children. Binary trees find applications in language parsing/representation of mathematical and logical expressions, finding duplicates in a given list of numbers. Binary search trees can be used for sorting a list of given numbers.
- *Depth-First Spanning Tree*: An edge (v, w) that leads to the discovery of an unvisited vertex during a depth-first search is referred to as a tree edge of a graph G. Collectively the tree edges of G form a depth-first spanning tree of G. DFS tree can be used to obtain a topological sorting, to find out the connectedness on a graph and to compute a spanning forest of graph, a cycle in graph and also bi-connected component, if any.
- *Breadth-First Spanning Tree*: BFS algorithm is applied to determine if a graph is bipartite, testing whether a graph is connected, computing a spanning forest of a graph, computing a cycle in a graph or reporting that no such cycle exists, etc. Also used to find the diameter of a tree having applications in network communication systems.
- *Spanning Tree Protocol*: A link layer network protocol that ensures a loop-free topology for any bridged LAN. Also allows a network design to include spare (redundant) links to provide automatic backup paths. *Multiple Spanning Tree Protocol*: Used to develop the usefulness of Virtual LANs (VLANs) further.
- *Broadcast and Peer-to-peer Networks*: Broadcast operation is fundamental to distributive computing. *Flooding Algorithms* construct a spanning tree which is

used for convergecast. Convergecast is collecting information upwards from the spanning tree after a broadcast. A broadcast spanning tree can be built such that each non-leaf peer forwards broadcast messages to its children along the tree.

- *Link Failure Simulation in Network*: One of the most important tools in network capacity planning is the ability to simulate the physical link failure scenario, where the planner would like to cut-off a physical link logically and then try to see how the network behaves due to that failure. During this process the planners would also try to identify all possible alternate routes between the nodes where they have logically torn down the direct physical link, so that they can distribute the entire traffic from that link to the existing path and see whether those links have enough capacity to handle additional load that has been generated due to the physical link failure simulation. Since the above-described process needs to be repeated for all the nodes in the network and hence the necessity of finding all possible spanning trees for the network topology is evident.

7 Conclusion

We have given here a new approach towards solving the well-known tree generation problem using the divide-and-conquer technique used in algorithms. The algorithm formulated in this paper is capable of computing all possible spanning trees of a simple, symmetric, and connected graph. The given graph has been divided into a number of partitions, which can be joined by a set of connectors. The selection of different connectors and the way they are being combined with the different partitions give rise to different spanning trees of the graph. Our algorithm does not generate any duplicate tree and also minimizes the formation of the circuit in its tree generation procedure, which has also been taken care of eventually. Till now, we have executed the algorithm on graph instances whose order is at most 22. Currently, we are working on larger graph instances compared to the ones considered in this paper to augment the results published in Table 1. Due to the limitation in a computing environment, some of the algorithms considered in this paper are taking time in weeks for execution.

References

1. Cormen, T.H., Leiserson, C.E., Rivest, R.L., Stein, C.: Divide-and-Conquer: Introduction to Algorithms, 3rd edn. The MIT Press, Cambridge, Massachusetts (2009)
2. Berger, I.: The Enumeration of Trees without Duplication. IEEE Trans. Circuit Theor. **14**(4), 417–418 (1967)
3. Char, J.P.: Generation of trees, two-trees, and storage of master forests. IEEE Trans. Circuit Theor. **15**(3), 228–238 (1968)

4. Gabow, H.N., Myers, E.W.: Finding all spanning trees of directed and undirected graphs. SIAM J. Comput. **7**(3), 280–287 (1978)
5. McElroy, M.D.: Algorithm 354: generator of spanning trees [H]. Commun. ACM **12**(9), 511 (1969)
6. Naskar, S., Basuli, K., Sen Sarma, S.: Generation of all spanning trees of a simple, symmetric, connected graph. In: National Seminar on Optimization Technique, Department of Applied Mathematics, University of Calcutta, p. 27 (2007)
7. Naskar, S., Basuli, K., Sen Sarma, S.: Generation of All Spanning Trees. Social Science Research Network (2009)
8. Naskar, S., Basuli, K., Sen Sarma, S.: Generation of all spanning trees in the limelight. In: Advances in Computer Science and Information Technology, Second International Conference, CCSIT 2012, Bangalore, vol. 86, pp. 188–192. Proceedings Part III published by Springer (2012)
9. Piekarski, M.: Listing of all possible trees of a linear graph. IEEE Trans. Circuit Theor. **12**(1), 124–125 (1965)
10. Sen Sarma, S., Rakshit, A., Sen, R.K., Choudhury, A.K.: An efficient tree generation algorithm. J. Inst. Electron. Telecommun. Eng. **27**(3), 105–109 (1981)
11. Trent, H.M.: Note on the enumeration and listing of all possible trees in a connected linear graph. In: Proceedings of the National Academy of Sciences. USA.40, pp. 1004 (1954)
12. Cherkasskii, B.V.: New algorithm for generation of spanning trees. Cybern. Syst. Anal. **23**(1), 107–113 (1987)
13. Hakimi, S.L.: On trees of a graph and their generation. J. Franklin Inst. **272**(5), 347–359 (1961)
14. Kapoor, S., Ramesh, H.: Algorithms for enumerating all spanning trees of undirected and weighted graphs. SIAM J. Comput. **24**(2), 247–265 (1995)
15. Matsui, T.: An algorithm for finding all the spanning trees in undirected graphs. In: METR93-08, Department of Mathematical Engineering and Information Physics, Faculty of Engineering, University of Tokyo. 16, pp. 237–252 (1993)
16. Matsui, T.: A flexible algorithm for generating all the spanning trees in undirected graphs. Algorithmica **18**(4), 530–543 (1997)
17. Mayeda, W., Seshu, S.: Generation of trees without duplications. IEEE Trans. Circuit Theor. **12**(2), 181–185 (1965)
18. Shioura, A., Tamura, A.: Efficiently scanning all spanning trees of an undirected graph. In: Research Report: B-270, Department of Information Sciences, Tokyo Institute of Technology, Tokyo. (1993)
19. Shioura, A., Tamura, A., Uno, T.: An optimal algorithm for scanning all spanning trees of undirected graphs. SIAM J. Comput. **26**(3), 678–692 (1997)
20. Minty, G.: A simple algorithm for listing all the trees of a graph. IEEE Trans. Circuit Theor. **12**(1), 120-120 (1965)
21. Smith, M.J.: Generating spanning trees. In: MS Thesis, Department of Computer Science, University of Victoria (1997)
22. Winter, P.: An algorithm for the enumeration of spanning trees. BIT Numer. Math. **26**(1), 44–62 (1986)

Circuit Synthesis of Marked Clique Problem using Quantum Walk

**Arpita Sanyal(Bhaduri), Amit Saha, Bipulan Gain
and Sudhindu Bikash Mandal**

Abstract Many applications such as Element Distinctness, Triangle Finding, Boolean Satisfiability, Marked Subgraph problem can be solved by Quantum Walk, which is the quantum version of classical random walk without intermediate measurement. To design a quantum circuit for a given quantum algorithm, which involves Quantum Walk search, we need to define an oracle circuit specific to the given algorithm and the diffusion operator for amplification of the desired quantum state. In this paper, we propose a quantum circuit implementation for the oracle of the marked clique problem based on Quantum Walk approach. To the best of our knowledge, this is a first of its kind approach in regards to the quantum circuit synthesis of the marked clique oracle in binary quantum domain. We have performed the simulation of the proposed oracle circuit in Matlab.

Keywords Quantum computing · Quantum walk · Marked clique finding

1 Introduction

Quantum computing has brought a new era in computing technology which is based on quantum theory. Quantum theory explains energy and matter in quantum level. While a classical computer works in only two state which is called bit, a quantum computer can work in more than two bit level that is superposition of states called qubit. Quantum Algorithm uses the power of Quantum Computing for solving various complex problems in more efficient way. Quantum algorithm is a step by step procedure where each step can be performed on a quantum register. Quantum algorithm replaces classical algorithm as it works faster than classical algorithm for some

A. Sanyal(Bhaduri)
Maharani Kasiswari College, University of Calcutta, Calcutta, India

A. Saha · B. Gain(✉) · S.B. Mandal
A.K Choudhury School of Information Technology, University of Calcutta,
Calcutta, India
e-mail: bipulan@gmail.com

© Springer Nature Singapore Pte Ltd. 2017
R. Chaki et al. (eds.), *Advanced Computing and Systems for Security*,
Advances in Intelligent Systems and Computing 567,
DOI 10.1007/978-981-10-3409-1_3

problems. For problems like factoring Shor's algorithm [1], for searching an unstructured database Grover's algorithm [2] performs best than any kind of classical algorithm.

Then comes the concept of Quantum Walk, which also gives quadratic speed up over classical algorithm for some practical problem. While a Grover's Algorithm gives complexity for finding a item in time $O(N^{\frac{1}{2}})$, Quantum walk can also search an item in $O(\sqrt{|N|})$ time but in case of grid, Grover's algorithm takes $O(N)$ while Quantum Walk takes $O(\sqrt{|N|})$. In classical random walk the particle moves one position either left or right depending on the flip of a fair coin. Such classical random walk can be on finite or infinite graph [3]. Quantum Walk is the quantum variation of Random walk in which, the particle moves in left and right with superposition of states. That is in classical Random walk if P+ and P– be the outcome of two probabilities of head and tail. Then in Classical Random walk these two values are same. If a quantum coin is tossed the coin can be in a superposition of two states. That is if the basis states are $|+\rangle$ and $|-\rangle$ then the state will be

$$|\psi\rangle = a|+\rangle + b|-\rangle \tag{1}$$

In classical random walk, a particle moves one step towards right with the probability p+ if outcome of the coin toss is a head and moves one step left with the probability P– if outcome of the coin toss is a tail but, in quantum walk the particles moves in left and right simultaneously with amplitudes a and b. We did not measure the position of the particle every time before the move in a quantum walk because this measurement or observation will lead the quantum walk into an ordinary classical walk. So quantum walk proceeds in the absence of any measurement or observation [4]. While in Classical Random Walk after n steps distance covered is $\sqrt{n}$ in Quantum Random walk after n steps total distance covered is n.

There can be two kind of Quantum Walk which are Discrete Quantum Walk and Continuous Quantum Walk. Discrete Quantum walk is performed by coin tossing followed by a shift operation where in continuous random walk no coin tossing is required, the walk is represented by a transition matrix. Difference between Classical Random Walk and Quantum Walk can be represented by the probability distribution graph [5] (Fig. 1).

A clique is a subset of a graph in which all vertices are connected. There can be more than one clique in a graph. Our aim is to find the marked clique in the graph. There are several existing algorithm for finding clique of a graph in classical computing. In 1973 Kerbosch et al. gave Bron Kerbosch algorithm for finding clique of a graph which has complexity of $O(3^{\frac{n}{3}})$ [6]. Later Patric Ostergerd [7] gave a fast algorithm over the existing algorithm which is a NP hard problem. Ashay Dharwadker gave another clique finding algorithm which is Polynomial Time Algorithm [8]. In Quantum Computing Alan Bojic gave a quantum algorithm with complexity $O(|V|\sqrt{2^{|V|}})$ in worst case and $O(\sqrt{2^{|V|}})$ in best case [9]. Mark Hilery et al. gave a procedure for finding marked subgraph of a graph using Quantum Walk [10]. They used Scattering Quantum Walk and used Grover's algorithm for

Fig. 1 Probability
distribution graph for
classical and random walk

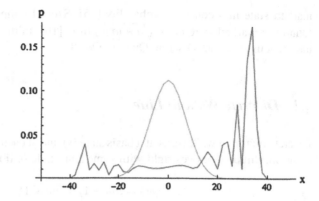

this process. But, there is no research work till now on circuit synthesis of finding
Marked Clique of a graph. The algorithm that we have proposed is totally based on
Discrete Quantum walk. Our algorithm takes only quadratic speed up compared to
its classical counterpart for discrete random walk. All other functions related to our
algorithm take linear time. We have used two coin operator here for marked and for
unmarked states. We have given circuits for each phases along with detailed expla-
nation about every Oracle used in the algorithm.

Our key contributions in this paper can be summarized as:

- Proposal of an algorithm that find a marked clique by using Discrete Quantum
walk.
- Circuit synthesis for marked clique.

For the execution of above steps we organize our paper as follows. In the Sect. 2,
we have discussed about several research walk that has been proposed by several scientist. In the subsection we elaborate the concept of discrete and continuous quantum
walk and depict its impact on a graph. In Sect. 3, we have given the detailed quantum
algorithm for finding Marked Clique. In Sect. 4, we have designed circuits and have
given detailed description of the circuit. In Sect. 5, simulation of our algorithm is
given. In Sect. 6 we have described the circuit for a graph of n-nodes. Concluding
remarks appear in Sect. 7.

2 Background

In 1993, **Y. Aharonov L. Davidovich and N. Zagury** first brought the idea of Quantum Walk [11]. After that several scientists discovered several quantum algorithm.
Ambainis was first to solve a natural problem that is element distinctness problem
which tests if every element in the set is distinct or not [12]. Then several quantum
algorithm Triangle Finding by **Magniez et al.** [13], Matrix Product Verification [14]
by Harry Buhrman, Robert Spalek has been discovered, **Magniez et al.** found the

marked state in a constant probability [15]. Szegedy improved over the work and found the marked state in $O(\sqrt{n * logn})$ time [16]. In the following subsection we have discussed about Discrete Quantum Walk.

2.1 Discrete Walk in Line

Discrete quantum walk starts at a basis state $|s\rangle$ and at each time it moves to left with some amplitude a, moves right with some amplitude b. If the basis state be $|n\rangle$

$$|\psi\rangle = a|n-1\rangle + b|n+1\rangle \tag{2}$$

This movement to the left or right depends on the outcome of the coin flip operation. Discrete Quantum Walk is the combination of coin flip and shift operation(SC).

Choice of coin is Hadamard coin $\begin{pmatrix} \frac{1}{\sqrt{2}} & \frac{1}{\sqrt{2}} \\ \frac{1}{\sqrt{2}} & -\frac{1}{\sqrt{2}} \end{pmatrix}$.

Coin flip decide in which state the particle will move from the current state and shift operation shifts the particle from the current state to next state. At each step we do two operations (Fig. 2).

Coin Flip Operation:
If the base state is $|n, 0\rangle$ then after flipping Hadamard coin the state will be.

$$|\psi\rangle = \frac{1}{\sqrt{2}}|n, 0\rangle + \frac{1}{\sqrt{2}}|n, 1\rangle \tag{3}$$

If the base state is $|n, 1\rangle$ then after flipping Hadamard coin the state will be

$$|\psi\rangle = \frac{1}{\sqrt{2}}|n, 0\rangle - \frac{1}{\sqrt{2}}|n, 1\rangle \tag{4}$$

That is the particle moves to the left and right direction with the amplitude $\frac{1}{\sqrt{2}}$.

Fig. 2 Discrete quantum walk along a straight line

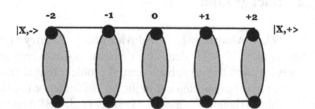

Shift Operation:

If the Current state is $|n, 0\rangle$ then

$$S|\psi\rangle = |n - 1, 0\rangle \tag{5}$$

and if current state is $|n, 1\rangle$ then

$$S|\psi\rangle = |n + 1, 1\rangle \tag{6}$$

Let us consider three steps of quantum walk. That is flipping the coin and shifting three times. If the starting state is $|0, 0\rangle \, |0, 0\rangle \rightarrow \frac{1}{\sqrt{2}}|0, 0\rangle + \frac{1}{\sqrt{2}}|0, 1\rangle$

$$\rightarrow \frac{1}{\sqrt{2}}|-1, 0\rangle + \frac{1}{\sqrt{2}}|1, 1\rangle$$

$$\rightarrow \frac{1}{2}|-1, 0\rangle + \frac{1}{2}|-1, 1\rangle + \frac{1}{2}|1, 0\rangle - \frac{1}{2}|1, 1\rangle$$

$$\rightarrow \frac{1}{2}|-2, 0\rangle + \frac{1}{2}|0, 1\rangle + \frac{1}{2}|0, 0\rangle - \frac{1}{2}|2, 1\rangle$$

$$\rightarrow \frac{1}{2\sqrt{2}}|-2, 0\rangle + \frac{1}{2\sqrt{2}}|-2, 1\rangle + \frac{1}{\sqrt{2}}|0, 0\rangle - \frac{1}{2\sqrt{2}}|2, 0\rangle + \frac{1}{2\sqrt{2}}|2, 1\rangle$$

$$\rightarrow \frac{1}{2\sqrt{2}}|-3, 0\rangle + \frac{1}{2\sqrt{2}}|-1, 1\rangle + \frac{1}{\sqrt{2}}|-1, 0\rangle - \frac{1}{2\sqrt{2}}|-1, 0\rangle + \frac{1}{2\sqrt{2}}|3, 1\rangle$$

In the above walk, after second step the particle is in state -2 and 2 with amplitude $\frac{1}{4}$ and is in state 0 with probability $\frac{1}{2}$. Up to this step the walk is a classical walk but after second step we can see that the particle moves along the left with all its amplitude whereas in classical walk the particle moves towards left and right with equal amplitude. This is quantum interference. If we consider more steps the particle moves towards more left side.

This movement of particle towards left side is because we choose a biased coin which moves from the state $|n, 0\rangle$ to $|n, 0\rangle$ with amplitude $\frac{1}{\sqrt{2}}$ and it moves from the state $|n, 1\rangle$ to $|n, 1\rangle$ with amplitude $-\frac{1}{\sqrt{2}}$. So we use a symmetric coin $\begin{pmatrix} \frac{1}{\sqrt{2}} & \frac{i}{\sqrt{2}} \\ \frac{i}{\sqrt{2}} & \frac{1}{\sqrt{2}} \end{pmatrix}$.

2.2 Continuous Quantum Walk

Continuous quantum walk was first proposed by Farhi and Gutmann. In continuous quantum walk, coin tossing is not necessary. The walk is defined by a transition matrix. The transition matrix which is hamiltonian (H) which will generate the evolution $U(t)$

$$U = e^{-IHT} \tag{7}$$

H is the Transition Matrix. If the continuous walk be on a graph. Then H is defined by $H_{ij} = -1$ if $i \neq j$ if i and j is connected by an edge and $H_{ii} = d_i$ if vertex i has degree d_i. In the next section, we will give a brief overview about discrete quantum walk on a graph.

2.3 Discrete Quantum Walk in a Graph

Our algorithm for finding Marked Clique is based on Discrete Quantum Walk on a graph, so here we discuss about Discrete Quantum Walk on a graph. [17] Let G(V, E) be a graph. V be the set of vertices and E be the set of edges. Each vertex v_i of the set V has a set of nodes which is the degree of the node. There is a coin register $|C\rangle$ and a state register $|v_i\rangle$ corresponding to ith vertex of the graph. If d is the degree of the vertices and n is the number of vertices then $log_2 d$ and $log_2 n$ qubits are needed for the $|C\rangle$ and $|v_i\rangle$ registers, One step of quantum walk on the graph consists of the following:

- For each vertex we do coin flip i.e. apply coin operator which acts on the coin register C which determines to which quantum state the vertex will move.
- Shift the vertex state from one state to another vertex state to which it is connected.

In the next section we have described the algorithm for finding Marked Clique for a graph.

3 Algorithm for Finding the Marked Clique

For finding marked clique, we need adjacency matrix first which, will be the input. In this Algorithm **Adj**[][] is the adjacency matrix of the graph, m is the number of rows and n is the number of columns of the matrix. **N** is the total number of nodes. V is the vertex set of the graph. C is the coin states of the graph. In the variable deg we have stored maximum degree of the graph. C' is the coin operator. For unmarked state C_0 coin is used and for marked state C_1 coin is used. If X_0 be any arbitrary vertex that is marked then

$$C' = I \otimes C_0 + |X_0\rangle\langle X_0| \otimes (C_1 - C_0) \qquad (8)$$

$C_0 = G$, G is Grover's coin operator; $C_1 = -I$ We have subset **N(v)** is the Neighbour set of each vertices for graph. *Marked* is the register where marked clique are kept. **connectvtx** is the function to check connectivity between the edges and if clique is found, it is kept in the *MRKVTX* register. **S** is the shift operator. The algorithm for finding marked clique is given below:

Algorithm 1: Finding Marked Clique

```
 1: INPUT : Adjacency Matrix of the Graph
 2: m ← row
 3: n ← column
 4: deg ← 0
 5: for i ← 1 to m do
 6:    d ← 0
 7:    for j ← 1 to n do
 8:       if adj[i][j] ≠ 0 then
 9:          d ← d + 1
10:       end if
11:    end for
12:    if deg <d then
13:       deg ← d
14:    end if
15: end for
```

16: $MRKVTX \leftarrow register([0, 0, ..N])$

17: $marked \leftarrow register([0, 0, ..N])$

18: $V \xrightarrow{H} \frac{1}{\sqrt{2}^n} \sum_{v \in (0,1)^2} |v\rangle$ where [H is the Hadamard operator]

19: $C \xrightarrow{H} \frac{1}{\sqrt{2}^n} \sum_{c \in (0,1)^2} |c\rangle$ where [H is the Hadamard operator]

```
20: for k ← 1 to √N do
21:    U ← SC' where [C' is coin operator and S is shift operator]
22:    UPDATE N(V) Neighbour set of corresponding vertex is updated
23: end for
24: for each v ∈ V do
25:    MRKVTX_v =FUNCTION connectvtx(v,N(v))
26: end for
27: marked = MRKVTX_v
28: marked ← C' [where C' is Coin operator]
```

3.1 Steps for Finding Marked Clique Using Our Proposed Algorithm

In the following subsections we will show the methodology for finding maximum clique of a graph using our proposed algorithm and with the help of an example.

3.2 Calculation of the Degree

We work on the graph given in the Fig. 3. At first we calculate the degree of each vertex. We traverse the adjacency matrix and test for non zero entries in the matrix corresponding to each vertex. After traversing the whole matrix we get maximum degree. According to the maximum degree calculate the qubits required to represent

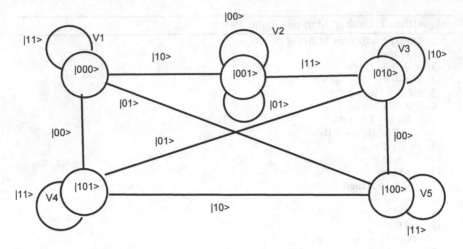

Fig. 3 Quantum walk along a graph

the coin states which is our C register. Total number of qubits required represent vertex states depend upon total number of vertices.

3.3 Applying Hadamard Gate

We apply Hadamard gate to vertex state and coin state so that we get all possible combination of vertex and coin. Then we define the adjacency matrix for the graph.

3.4 Starting Quantum Walk

We start quantum walk from an arbitrary vertex. Quantum walk start from a vertex

choosen randomly. We uses a Hadamard coin $\begin{pmatrix} \frac{1}{\sqrt{2}} & \frac{1}{\sqrt{2}} \\ \frac{1}{\sqrt{2}} & -\frac{1}{\sqrt{2}} \end{pmatrix}$. After tossing coin the ver-

tex moves to all its neighbouring vertex which is the action of shift operator. For example if we perform quantum walk in our example graph. We have maximum degree which is 4 here. So, we need 2 qubits to represent the coin states. As we have 5 vertices so we need 3 qubits to represent vertex state. So if we start from the vertex V1 then the walk proceeds like

$$SC|00\rangle|000\rangle = \frac{1}{2}S[|00\rangle|000\rangle + |01\rangle|000\rangle + |10\rangle|000\rangle + |11\rangle|000\rangle]$$
$$= \frac{1}{2}[|00\rangle|101\rangle + |01\rangle|100\rangle + |10\rangle|001\rangle + |11\rangle|000\rangle] \tag{9}$$

If we start from the vertex V2 then the walk proceeds like

$$SC|00\rangle|001\rangle = \frac{1}{2}S[|00\rangle|001\rangle + |01\rangle|001\rangle + |10\rangle|001\rangle + |11\rangle|001\rangle]$$
$$= \frac{1}{2}[|00\rangle|001\rangle + |01\rangle|001\rangle + |10\rangle|000\rangle + |11\rangle|010\rangle] \tag{10}$$

Neighbour set of the vertices are updated with this quantum walk.

3.5 Checking Interconnection Between Vertices

After discrete quantum walk on the graph we have separate subset of adjacent edges for each vertex. For each subset interconnection between adjacent vertices are checked using adjacency matrix. We have an edge oracle which, check for the interconnection of the adjacent vertices. For each vertex we execute the oracle. Based on the outcome our function update MRKVTX register. After executing the oracle for each vertex we may get one or many clique.

3.6 Getting the Clique

We can have multiple clique of the same graph. So, after executing the algorithm we find whether there is a marked clique or not.

3.7 Coin Operator on Marked List

After getting the final list of marked vertex we apply coin operator on the marked states. Coin operator acts as a diffusion operator inverses the state and amplify the amplitude of the marked state. Diffusion operator is Hadamard transform

H xor followed by phase flip operation $D_m = \begin{pmatrix} -1+\frac{2}{d} & \frac{2}{d} & \cdots & \frac{2}{d} \\ \frac{2}{d} & -1+\frac{2}{d} & \cdots & \frac{2}{d} \\ \cdots & \cdots & \cdots & \cdots \\ \frac{2}{d} & \frac{2}{d} & \cdots & -1+\frac{2}{d} \end{pmatrix}$

[2]. It acts only on the marked vertex and increase its amplitude than any other vertex of the graph. Thus we get the amplified state of marked clique.

4 Analysis of the Circuit

4.1 Applying Hadamard Gate

We are making circuit for our graph given in Fig. 3. Here we have 5 vertices. So total 3 qubits are needed to represent the vertex state. Maximum degree of the graph is 4. So, we need 2 qubits to represent the coin states. The coin states will be $|00\rangle$, $|01\rangle$, $|10\rangle$, $|11\rangle$. So we need 4 circuits to represent these four coin states. We initialize the vertex to the initial state $|00\rangle$ and apply Hadamard gate so that we get all possible combination of the vertices.

4.2 Circuit for Different Coin States

The circuit for coin state $|00\rangle$ is presented in Fig. 4a, we have 12 gate levels here. At first we have identified which state has a transition to which states and have drawn the circuit accordingly with the help of CNOT, NOT, C2NOT gate.

The circuit for coin state $|01\rangle$ is presented in Fig. 4b. Here we have used 7 gate levels. There are $|000\rangle$, $|010\rangle$, $|100\rangle$, $|101\rangle$ states which has transition for coin state $|01\rangle$. We have used C2NOT, NOT and CNOT gate for the circuit.

Circuit for $|10\rangle$ coin State is presented in Fig. 4c. Only $|000\rangle$, $|001\rangle$, $|100\rangle$, $|101\rangle$ states are changed. We have 6 gate levels. Here the 1st qubit and the 2nd qubit of the vertex state is not changing at all so we keep it as it is. Only 3rd qubit inverts at first level. On the 2nd level 2nd qubit acts as a control bit and inverts the 3rd qubit if its value is 1. We have used only NOT and CNOT gate.

Circuit for $|11\rangle$ Coin State is presented in Fig. 4d. Here we have six gate levels. Only two states has been changed $|001\rangle$ and $|010\rangle$. All the other states remains the same. So we have used 3 CNOT gate and 3 C2NOT gate for this purpose [18].

4.3 Interconnection Between Adjacent Vertices

We have separated subsets of edges for each vertices. So, for each vertices of the graph we check the interconnection between its corresponding neighbor vertices using a quantum oracle. *connectvtx* is the function that performs this function. For each subset we find whether there is a marked clique or not. We have given quantum circuit for the oracle for checking interconnection between vertices. We have drawn

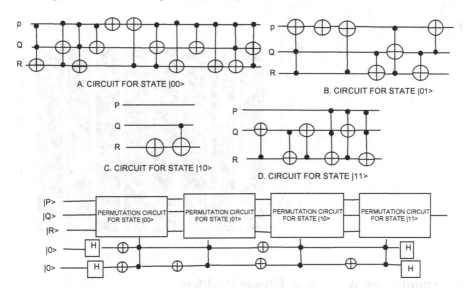

A. CIRCUIT FOR STATE |00>

B. CIRCUIT FOR STATE |01>

C. CIRCUIT FOR STATE |10>

D. CIRCUIT FOR STATE |11>

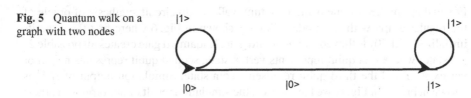

Fig. 4 Full circuit of all states

Fig. 5 Quantum walk on a graph with two nodes

the circuit for quantum walk for a graph with 2 nodes. Here we have only two nodes. So starting from $|0\rangle$ node if coin state is $|0\rangle$ shift operator moves the walker to state $|1\rangle$ otherwise walker remains in state $|0\rangle$. Graph is given in Fig. 5.

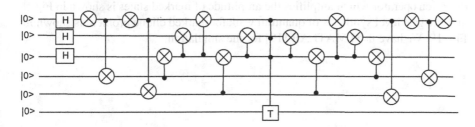

Fig. 6 Gate level representation of oracle for a graph with two nodes

A. Sanyal(Bhaduri) et al.

Fig. 7 Simulation output of $H^{\otimes 3}$

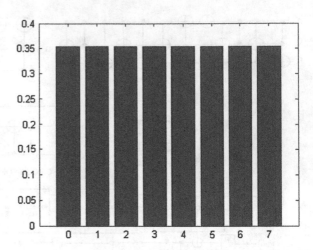

5   Simulation of Marked Clique Problem

For finding marked clique using quantum walk we do circuit synthesis of oracle of the regular graph with two nodes which is shown in Fig. 6 where first three qubits initialized with $|0\rangle$ followed by hadamard gate. Hadamard gate creates all possible 2^3 (8) states where first qubit represents vertex state, second qubit represents neighbor vertex state and the third qubit represents coin state. Simulation output of $H^{\otimes 3}$ is shown in Fig. 7. In Fig. 8, we have shown the simulation result of our proposed oracle where we get two marked states which are $|010\rangle$ and $|100\rangle$. From the first marked state we can easily state that vertex $|0\rangle$ is connected to neighbor vertex $|1\rangle$ by an edge and the probability distributed to vertex $|1\rangle$ from vertex $|0\rangle$ using coin state $|0\rangle$. In similar way the second marked state defines the same. With the help of these two marked state we easily conclude that there is a subgraph of two connected vertices which means the graph has clique 2. The output of the oracle acted upon by the diffusion operator which amplifies the amplitude of marked states is shown in Fig. 9 and the gate level synthesis of quantum walk for marked clique problem is shown in Fig. 10. We have used MATLAB [19] for the simulation.

Fig. 8 Simulation output of oracle

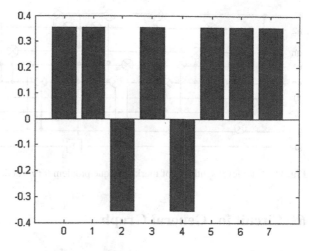

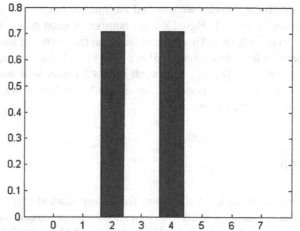

Fig. 9 Output after amplitude amplification

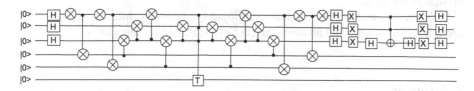

Fig. 10 Gate level synthesis of marked clique problem for two node graph

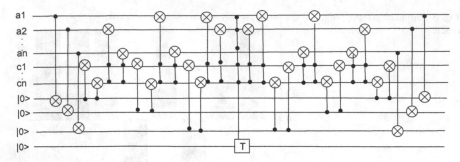

Fig. 11   Gate level synthesis of marked clique problem for n–node graph

6   Circuit for General Graph

Like the above graph for 2 nodes we can find the marked clique of a graph having n–number of nodes in Fig. 11. Based on the number of each node, we select the register size for vertex, which is $\lceil \log N \rceil$ and based on the degree of each vertex, we select the register size for coins, which is $\lceil \log C \rceil$. In Fig. 11, $\{a_1, a_2...a_n\}$ represents qubits for vertex state and $\{c_1..c_n\}$ represents qubits for coin state and rest of the qubits are used as ancilla qubit. However, for simulation in Matlab we have to limit our example to only 2 node graph.

7   Conclusion

In our paper, we have given total algorithm for finding marked clique of a regular graph using Discrete Quantum Walk. We proposed quantum circuit for the marked clique problem of a graph as an example application of Quantum Walk. We have designed a gate level synthesis of the oracle and also verified the circuit instances through simulation.

In future, we plan to develop quantum logic synthesis for different applications through this technique and analysis the cost of the respective algorithm.

References

1. Shor, F.W.: Polynomial-time algorithms for prime factorization and discrete logarithms on a quantum computer. SIAM J. Sci. Stat. Comput. **26**, 1484 (1997). arXiv:quant-ph/9508027
2. Grover, L.: A fast quantum mechanical algorithm for database search. In: Proceedings of STOC'96, pp. 212–219. quant-ph/9605043
3. Ambainis, A.: Quantum walks and their algorithmic applications. Int. J. Quant. Inf. **1**(4), 507–518 (2003). arXiv:quant-ph/0403120
4. Kempe, J.: Quantum Random Walks—an Introductory Overview. quant-ph/0303081

5. Childs, A.M., Farhi, E., Gutmann, S.: An Example of the Difference Between Quantum and Classical Random Walks. quant-ph/0103020
6. Bron, C., Kerbosch, J.: Algorithm 457: finding all cliques of an undirected graph. Commun. ACM (ACM) **16**(9), 575–577 (1973). doi:10.1145/362342.362367
7. Ostergerd, P.: Discrete Appl. Math. **120**, 195 (2002)
8. Dharwadker, A.: The Clique Algorithm. Amazon, pp. 48 (2011). ISBN:978-1466391215
9. Bojić, A.: Quantum Algorithm for Finding a Maximum Clique in an Undirected Graph. ISSN 1846-9418
10. Hillery, M., Reitzner, D., Buzek, V.: Searching via Walking: How to Find a Marked Subgraph of a Graph Using Quantum Walks. arXiv:0911.1102v1 [quant-ph]
11. Aharonov, Y., Davidovich, L., Zagury, N.: Quantum random walks. Phys. Rev. A **48**, 1687 (1993)
12. Ambainis A.: Quantum Walk Algorithm for Element Distinctness. quant-ph/0311001
13. Magniez, F., Santha, M., Szegedy, M.: An O(n~1.3) Quantum Algorithm for the Triangle Problem. quant-ph/0310134
14. Quantum Verification of Matrix Products. arXiv.org/pdf/quantph/0409035
15. Krovi, H., Magniez, F., Ozols, M., Roland, J.: Quantum Walks Can Find a Marked Element on any Graph. quant-ph/1002.2419
16. Szegedy, M.: Quantum speedup of markov chain based algorithms. In: Proceedings of the 45th Annual IEEE Symposium on Foundations of Computer Science (FOCS'04) 0272-5428/04
17. Aharonov, D., Ambainis, A., Kempe, J., Vazirani, U.: Quantum Walks on Graphs. arXiv:quant-ph/0012090v2
18. Chakrabarti, A., Lin, CC., Jha, NK.: Design of quantum circuits for random walk algorithms. In: 2012 IEEE Computer Society Annual Symposium on VLSI
19. MATLAB, R2012b, The MathWorks. Natick, MA (2012)

Abort-Free STM: A Non-blocking Concurrency Control Approach Using Software Transactional Memory

Ammlan Ghosh, Rituparna Chaki and Nabendu Chaki

Abstract Software transactional memory (STM) is a promising approach for concurrency control in parallel computing environment. The non-blocking progress implementations for STM forces transactions to abort. Although this is primarily done to ensure block-freedom, it may lead to poor system performance. This paper proposes a new Abort-Free STM methodology (AFTM) to achieve abort-free execution so that a group of processes, which are contending for a common set of concurrent objects can commit in finite number of steps. The proposed STM allows wait-free, non-blocking execution of multiple read and write transactions on shared data object without aborting any of the transactions. The important properties of AFTM have been proved towards establishing its advantages.

Keywords Concurrency control · Software transactional memory · Obstruction freedom · Abort freedom

1 Introduction

Software Transactional memory (STM) [1] has emerged as an important research area for multi-core processor. STM allows a sequence of read/write operations on a sharable object in a concurrent environment. It ensures that the execution of transaction will be either successful, in which case transaction will commit or unsuccessful, in which case transaction will abort. If conflict between two

A. Ghosh (✉) · R. Chaki · N. Chaki
University of Calcutta, Kolkata, India
e-mail: ammlan.ghosh@gmail.com

R. Chaki
e-mail: rchaki@ieee.org

N. Chaki
e-mail: nabendu@ieee.org

© Springer Nature Singapore Pte Ltd. 2017
R. Chaki et al. (eds.), *Advanced Computing and Systems for Security*,
Advances in Intelligent Systems and Computing 567,
DOI 10.1007/978-981-10-3409-1_4

53

transactions occurs, when both of them try to access same sharable data object, STM aborts one of the transactions to ensure consistency. When a transaction is aborted, all its effects are discarded and it tries to execute again at later time and commits eventually. Frequent aborts lead to a devastating effect on the overall performance of the system [2]. Reducing number of aborts is one of the biggest challenges that recent researches in STM are emphasizing to achieve good performance and guarantees parallelism.

Permissiveness is an aspect of progress that does not abort any transaction unless it is necessary to ensure consistency [3]. The permissiveness can be single-version or multi-version and both of them are lock-based. Single-versioned permissiveness [4] is a weaker condition and suffers from various spurious aborts. Conversely multi-versioned permissiveness (MV-Permissiveness) [5, 6] ensures complete abort-freedom for read-only transactions and permits write transactions to abort only in case of data inconsistency.

A few recent researches [7–9] have worked on progressiveness by avoiding conflicts among transactions. Lazy Snapshot Algorithm (LSA) [7] is a time based algorithm that uses a discrete logical global clock. The algorithm uses Bloom Filter to reduce number of false conflicts. LSA has significant improvement in STM progressiveness. However, it allows aborting of a write transaction when the transaction works with the older version of the data object. Aydonat and Abdelrahman used conflict-serializability model of database system into STM to reduce rate of transactional abort when transactions try to access shared-data in presence of contention [8]. The throughput of this system depends on the efficiency and appropriateness of conflict-serializability technique. S. Dolev, et al. propose SemanticTM [9] that uses t-var, a list of instructions for each t-variable to achieve non-blocking abort-free transaction execution. The drawback of the system is that the transaction must know the set of t-variable in advance. Abort-freedom for write transactions is presented in [10], where a transaction is allowed to proceed immediately without affecting the execution of other concurrently executing transaction in presence of contention. This system doesn't discuss the scenarios where multiple transactions try to access the same data objects for update. A wait-free, non-blocking implementation of STM is also presented in [11]. It aims to allow multiple write transactions, updating the same transactional memory object, without aborting any of the transactions. The throughput of this method deteriorates severely, when the length of the cascading chain becomes larger. Besides, the solution allows aborting of transactions for certain conditions.

The major contribution of this paper is to propose a new Abort-free Software Transactional Memory (AFTM). The proposed algorithm ensures that a group of transactions accessing a common shared data object are executed in a non-blocking, abort-free manner such that all the transactions are eventually completed in a finite number of steps. It is assumed that every transaction within the group executes in finite steps as well. Thus, the proposed algorithm offers a wait-free non-blocking solution that promises high throughput by avoiding abort operation to negotiate access conflict between transactions.

This paper is structured as follows. First, we describe the new approach for Abort-free STM implementation. Second, we present the critical analysis of the proposed model. Third, the performance evaluation result is included. The paper ends with concluding remarks.

2 Proposed Abort Free STM

2.1 Terminologies

In this section, some of the important terminologies that have been used in the rest of the paper are defined to clearly identify the premises under which the work has been proposed.

Set of Transactions: A set of n transactions $T = T_1, T_2,... T_n$ is a collection of finite length transactions that are uniquely identified with the subscript value in the order of their occurrences. The set T and its constituent transaction are assumed to follow the properties mentioned below:

- Individual transactions in T are either for read only operation, or for write operation.
- T_k follows T_m $\forall$ k, m $\in$ [1..n] iff transaction T_k occurs after transaction T_m.
- T_k precedes T_m $\forall$ k, m $\in$ [1..n] iff transaction T_k occurs before transaction T_m.
- T_k and T_m have distinct order of occurrences in T for any values of k, m $\in$ [1..n] i.e., either T_k follows T_m or T_k precedes T_m.
- An appropriate process synchronization mechanism must ensure that set T is completed in finite time.

Read-data Inconsistency: Let T_k $\in$ T and X be a sharable data object. During its occurrence T_k reads value of X and stores it as X'. At commit point of T_k, if $X = X'$ then read data is consistent; otherwise read data is inconsistent.

Dirty Reading: Dirty reading, as defined in the literature [12], occurs in a transaction execution history when a transaction, say T_1 modifies a data item and before it commits or aborts, some other transaction, say T_2, reads that data item. This will result in an inconsistency if transaction T_1 aborts, T_2 has read a data item that was never committed.

Atomicity: Either the whole transaction is executed (when it successfully commits) or none (when it aborts). This property is often referred to as the all or nothing property [12].

Consistency: Every transaction starts from a consistent view of the state and leaves the system in another consistent state, provided that the transactions would do so if executed sequentially [12].

Isolation: Individual memory updates within an ongoing transaction are not visible outside of the transaction. All memory updates are made visible to the rest of the system only when the transaction commits [12].

Abort: In non-blocking synchronization, abort implies that a transaction has detected an inconsistency and must undo its effects i.e. all the changes performed by that transaction on shared objects are completely discarded. The resources are released and the transaction tries to execute later.

Abort Freedom: We say that an STM implementation for set of transactions T is Abort-Free (AFTM) if for every transaction $T_i \in T$, transaction T_i is never forcefully aborted.

Wait or Die: In the existing TM approaches [13, 14], a transaction either has to die in presence of contention, may be after random time of wait, before the contention manager decides which of the transactions to abort. In AFTM, a novel approach is proposed such that neither a transaction has to wait at its commit point, nor any of the two or more transactions in contention is to be aborted.

Transaction Aborting versus Re-execution: When a transaction aborts, it undoes all its effect and tries to execute again at later time. When transaction needs to re-execute afresh, resources are freshly allocated and the transaction is newly scheduled. This leads to unpredictable performance and reduced concurrency. The term Re-execution, used in this paper, implies that instead of aborting and executing later, transaction will re-execute immediately in case of inconsistency. In the proposed Re-execution process, a transaction reads the shared object for which it faces inconsistency and re-executes its operations. This method is able to avoid the threat towards maintaining the serializability issues in case of aborting a transaction.

As for example, say, K number of concurrent transactions are identified and scheduled for execution. Let any one of these K transactions, say A, is aborted at time T. The remaining (K − 1) transactions eventually will complete. By this time, more transactions may have been added in the job-queue. These newer transactions are not necessarily concurrent with A. Thus, scheduling the aborted transaction A, at a later instance, could be a threat to serializability, depending on when it is scheduled again.

2.2 Basic Concepts

The proposed AFTM algorithm is a non-blocking transactional memory [15, 16], which ensures that an entire set of transactions, both write and read-only, completes the execution in a finite time without aborting any transaction in the process. It differs from other non-blocking STMs in the sense that instead of aborting a transaction in presence of contention, the transaction is allowed to commit maintaining the progress condition. The efficiency of this algorithm lies on the judgment, whether a re-execution of operations for a transaction is efficient enough over the abort of a transaction. The primary motivation for abort is to avoid the occurrence of deadlock and to maintain the progress condition. When transaction aborts, it must undo its operation and re-execute later on. After an abort, when transaction re-executes afresh, the re-initiation of transaction, re-scheduling, re-allocation of shared objects are required, which reduces the throughput and increases the overhead.

All the transactions in the proposed system access the sharable data object, available globally, as per their occurrence. Whenever a transaction is ready to commit it checks the data consistency i.e., the data value it has read at the time of its initiation and the current data value. If data value is consistent then transaction commits without bothering the status of other transactions those are accessing this data object. Otherwise transaction re-executes its operation after reading the data object's current value and again tries to commit in the same way.

2.3 Data Structure

The data Structure for the proposed AFTM method is partially influenced by DSTM [16]. An object in AFTM is read and/or written in a transaction by returning a transactional memory object (AFTMObject) wrapper for the object. The code, that executes the transaction, opens the AFTMObject to point the original data and manipulates the data object directly.

The AFTMObject is depicted in Fig. 1a. It contains four fields: (a) TRN_INT[] is an array of transaction ids, where each id points to the corresponding transaction descriptor. (b) TRN_CMT[] is also an array of transaction ids, which is stored as per the commit order of the transactions. (c) DATA field contains the actual data object. The level of indirection for accessing the data has been minimized by storing data in-place as it is in [17]. (d) TRN_CNT stores the count of uncommitted transactions. When a transaction initiates, the value of TRN_CNT is incremented by one and whenever a transaction commits successfully the value is decremented by one.

Transaction Descriptor (Fig. 1b) contains two fields. (a) T_ID is a unique transaction identification number. Whenever a transaction is initiated, it gets a unique ID, which is stored into TRN_INT[] of AFTMObject. (b) ReadData filed is used to store the data object's last committed value once the descriptor is initiated.

AFTM uses two layers of memory regions i.e. public and non-public regions. AFTMObject resides in the public memory region. The transaction descriptor has

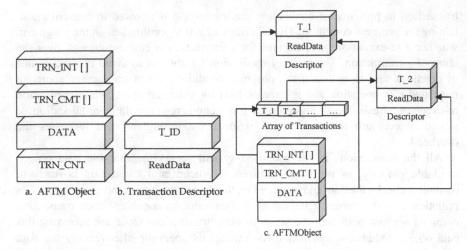

Fig. 1 a, b Data structure of Abort-Free Transactional Memory (AFTM), **c** Multiple transactions sharing data object concurrently by forming a chain

its own private memory region where it keeps the data object value that it reads. At commit point, a transaction checks the data consistency by comparing this value with the data object's the then current value. When a transaction commits successfully its private memory is being released.

2.4 Opening Data Object

This section describes data accessing method by the proposed AFTM algorithm that aims to ensure that a group of transactions using a common shared object are executed in a non-blocking, abort-free manner such that all the transactions are completed within finite number of steps. The logical flows of opening a data object for both read and write operations (shown in Fig. 1c) are as follows:

- Transaction T initiates and accesses an object for read or write operation. It gets a unique transaction id, which is stored in the T_ID field of its descriptor.
- T points to the TRN_INT[] field of AFTMObject. If TRN_INT[] contains NULL value, it means T is the first transaction that is opening the data object. Thus, the reference of T is inserted into the TRN_INT[] as a first element. Otherwise, its reference is inserted as the last element into the TRN_INT[].

- The value of the TRN CNT field of AFTMObject is incremented by one, as T is the newly initiated and not yet committed.
- Transaction T reads the data object value from DATA field of AFTMObject and stores it into its ReadData Field of Transaction Descriptor.
- At commit point T checks for the data consistency by comparing values of ReadData field and DATA field. T can commit only when value is consistent, otherwise T re-executes by reading DATA field's current value.
- When T commits successfully, the transaction id, T_ID is stored into the TRN_CMT[] as the last element to store the commit order of the transactions. If T is a write transaction then it writes the committed value into the DATA field.
- Transaction descriptor is released from the memory when T commits successfully.
- The value for TRN_CNT in AFTMObject is decremented by one as transaction T is successfully committed.

2.5 Committing of Transactions

Let us take an example, depicted in Fig. 3, to illustrate the abort-free execution in AFTM. A set of transactions, T_1, T_2, T_3, T_4 and T_5 are initiated one after another. Among them T_1, T_3 and T_5 are write transactions and T_2 and T_4 are read-only transactions. Transaction T_2 is initiated after T_1 but before T_1's commit. Thus, when T_2 reaches its commit point before T_1's commit, it finds a consistent data value and commits successfully. After T_2's commit, T_1 reaches its commit point and also finds consistent data as no other transaction has updated data value yet, thus T_1 also commits successfully.

Transaction T_3 is initiated after the T_1 and T_2. It has read the data object before the commit of write transaction T_1. Thus the probability of getting an inconsistent value for T_3, at the commit point, is quite high. Transaction T_4 is initiated after the last successful commit operation of updated transaction T_1 and no transaction is yet updated the data object before its commit point. Thus T_4 commits successfully. The scenario is same for the transaction T_5, which also commits in same manner. After commit of T_5, T_3 reaches it commit points. It is important to mention here that two write transactions, T_1 and T_5, as per the example, have been committed successfully in between respectively. As a result the data value that T_3 has read is now different i.e. inconsistent. Thus at the commit point, finding inconsistent data, T_3 reads the current data object value (latest is written by T_5) and re-executes its operation from there and successfully commit thereafter as per the given example.

It is important to discuss that how transaction T, which is re-executed may be for several times, will commit eventually in finite time. For a set of n transactions T_1, T_2,... T_n, each having a finite length, any transaction T_k eventually commits in finite time if none of the transactions are aborted in between. If any transaction T_k re-executes multiple times due to intermittent completion of several small write transactions, then T_k will be delayed by a finite time which is a function of the length of the small write transactions.

2.6 Algorithm

Algorithm 1 states the proposed algorithm. For all transaction $T_k \in T$ wants to access the data object; procedure Acquire is evoked and then T_k executes its operations by evoking procedure Execute (line 5, 6).

In the Acquire procedure (line 9 to 14) transaction T_k is included in the initiated transaction list i.e. TRN_INT. T_k reads data-object's value from DATA and stores it into ReadData field. As a new transaction is initiated the number of transaction count (TRN_CNT) is increased by 1.

Transaction T_k may be read-only or write. In case of write transaction it stores the modified value into some local variable (line 19) and tries to commit.

In the TryCommit process (line 23 to 31) T_k checks for the data consistency by comparing its ReadData field value with DATA (line 25). In case consistent data it executes commit process (line 32 to 40) otherwise it again retrieves the current value of DATA. If T_k is a Read-only transaction, it commits, otherwise, in case of write transaction T_k saves committed value from its local data i.e. tempDATA into the global data i.e. DATA and commits (line 34 to 36). After the successful commit of T_k, transaction's commit order is stored into the TRN_CMT[] (the array that stores the commit order of the transactions; line 37). The transaction count (TRN_CNT) is decremented by 1 (line 38) and T_k is removed from the initiated transactions' list (line 39).

Algorithm 1 Abort-Free Execution in AFTM

Initial State: A sequence of n transactions T_1, T_2, .., T_n that are active simultaneously and share the same data object stored in DATA field of AFTMObject. Transaction T_k is the kth transaction in the sequence.

Claims: Multiple transactions are allowed to execute both read and write operation on a sharable data object.   Each Transaction commits in finite time without aborting. No dirty reading takes place.

```
1.     procedure main(Tₖ)
2.     begin
3.     ∀ (Tₖ ∈ T)
4.     do
5.         Acquire(Tₖ);        /* Acquire AFTMObject data at start of execution for Tₖ */

6.         Execute(Tₖ);        /* Execution of transactional operation, read/write */
7.     end do
8.     end procedure

9.     procedure Acquire(Tₖ)
10.    begin
11.        TRN_INT[Element+1]← Tₖ
12.        Tₖ.ReadData←DATA;
13.        increase TRN_CNT by 1;
14.    end procedure

15.    procedure Execute(Tₖ)            /* Execute transactional operation */
16.    begin
17.        Execute steps of Tₖ;
18.        if Tₖ is Update Transaction then
19.              Store new data value in temporary variable temp_DATA;
20.        end if
21.        TryCommit(Tₖ);          /* When transaction is ready to commit */
22.    end procedure

23.    procedure TryCommit(Tₖ)
24.    begin
25.        if Tₖ.ReadData.Value = DATA then
26.              Commit(Tₖ);
27.        else    /* This is when DATA is different from ReadData.Value for Tₖ. */
28.              Tₖ.ReadData←DATA;
29.              Execute(Tₖ); /* Tₖ will be re-executed after freshly acquiring DATA */
30.        end if
31.    end procedure

32.    procedure Commit(Tₖ)
33.    begin
34.        if Tₖ is Update Transaction then
35.              Update TMObject Data with temp_DATA;
36.        end if
37.        TRN_CMT[] ←Tₖ.T_ID    /* Tₖ's commit order is saved into TRN_CMT[] */
38.        decrease TRN_CNT by 1;
39.        remove TRN_INT[Element=Tₖ]
40.    end procedure
```

---

3 Making AFTM Abort Free: A Critical Analysis

When two transactions concurrently access the same resource and at least any one
of them is involved in write, the traditional approaches of STM [14, 16, 17] allow
one of the transactions to commit while other is aborted. The proposed AFTM
ensures progress without any abort while ensuring progress condition. The algorithm allows any transaction to commit when the data value read at the time of
transaction's initiation is consistent with current value of the data object at its
commit point. If the data value is inconsistent then a transaction re-executes its
operation before trying to commit again.

Lemma 1 *AFTM ensures that execution of finite length transaction T consisting of
all finite-length transactions will be completed in finite number of steps.*

Proof Let's assume that the statement of Lemma 1 is false; i.e., AFTM cannot
complete a finite set of finite-length transactions in finite steps.

As per Lemma 1, if read-data is inconsistent at commit point then T_k re-executes
its operation after acquiring the current value of shared data object, say X. After
re-executing its operation when T_k invokes procedure TryCommit, it may again
find inconsistent data and re-execute once again.

According to our assumption this may repeat for an infinite period. However,
this is possible only if either set T has infinite number of transactions or one or more
members of transactions in T is of infinite length.

This is directly in contradiction with the statement of the Lemma in terms of the
definition of T.

Hence, the statement of Lemma 1 is proved by contradiction.

AFTM considers T as a finite set of transactions at a time instance t. If at any
time instance t' for all t' > t, some newer transactions are initiated, then these new
transactions are included in the transaction set to create T'. Here, T is a proper
subset of T'. Eventually, T' is also a finite set of transactions. Hence if Lemma 1 is
true for T, then it's also true for T'.

Lemma 2 *AFTM ensures wait-freedom*

Proof Wait freedom guarantees that all processes contending for a common set of
concurrent objects make progress in a finite number of their individual time steps
[13].

Suppose, by contradiction, a transaction T_i is invoked by a process P_i and never
returns. Thus transaction T_i has been aborted. As per the proposed algorithm, P_i
never stops executing within its transaction. Moreover by **Lemma** 1, every transaction completes their operations in finite time, which is a contradiction with the
assumption. Thus, proposed method ensures wait-freedom.

Lemma 3 *AFTM ensures data consistency in presence of contention*

Proof In AFTM, all the transactions $T_x \in T$ at their respective commit points, checks if the value of the AFTMobject with which it initiated its execution, is still same in the global data structure.

The consistency property of STM demands that every transaction starts from a consistent view of data object and leaves the system in another consistent view provided, as if, the transaction is executed sequentially. Let's assume that AFTM doesn't support the condition of consistency, i.e., starting from a consistent state, AFTM takes the system to an inconsistent state by executing a transaction. This could be checked in two alternate situations:

(i) Transaction $T_x \in T$ commits and this results in data inconsistency. Here, commit by T_x implies that no new transaction has committed after transaction T_x has accessed the AFTMobject and before T_x reaches its commit point. This again may be considered in two different contexts depending on whether T_x is a read-transaction or write-transaction. Let's assume that T_x is read-transaction and yet the state is inconsistent. This is absurd as T_x did not change the AFTMobject and the previous state of AFTMobject has been consistent by definition. Now consider that T_x is a write-transaction and the state is inconsistent. This also is absurd as the current write of AFTMobject does not follow any contention. No new transaction committed after T_x accessed the value of AFTMobject at the beginning of its execution. Thus if the previous state of AFTMobject is consistent, then current state cannot be inconsistent. Hence, the assumption that AFTM doesn't support consistency is absurd when the AFTMobject value remains unchanged at the commit point for any Transaction $T_x \in T$

(ii) Transaction T_x re-executes at the commit point of transaction T_x and leaves the state inconsistent. This too is absurd as before re-execution, neither T_x commits nor it writes back anything to AFTMobject. Hence, Lemma 3 is proved by contradiction.

Lemma 4 *AFTM design ensures Abort Freedom*

Proof As per Lemma 1 and Lemma 2 the proposed algorithm ensures system-wide progress. As per Lemma 1, a transaction T_i may re-execute its steps several times in case of data inconsistency. As $T_x \in T$, which is a finite set of finite-length transactions thus this re-execution process cannot be continued for infinite times. And moreover re-execution of T_i helps other transactions to commit successfully. This way every transaction is successfully brought to completion and hence the global progress for the system can be achieved. For these whole process no transactions, neither read-only nor write transaction is ever aborted by other transactions or by itself. Thus AFTM ensures abort-freedom.

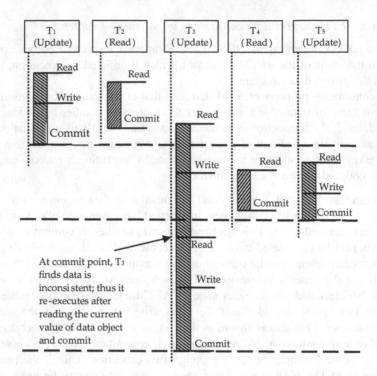

Fig. 2 Abort-free execution

Lemma 5 *AFTM is correct in terms of atomicity, isolation, and consistency*

Proof Every transaction in the proposed algorithm is atomic. AFTM has no separate requirements from Transactions and hence atomicity prevails by default. A transaction $T_x \in T$, does not access any intermediate value of shared object while being processed by another transaction $T_k \in T$. AFTMobject, as shown in Fig. 2, can access the data object that is updated by successfully committed transactions only. Therefore, data isolation property for the transactions is maintained in AFTM. As par Lemma 3, AFTM ensures atomicity. Hence, the proposed AFTM is correct in terms of atomicity, isolation and consistency.

4 Performance Evaluation by Experiments

We perform this prototype implementation on Intel Core i7, 64 bit processor with 8 GB memory and 2 MB L2 cache, running on Linux and using GCC 5.2 compiler. The efficiency and performance improvement of the proposed AFTM algorithm has been compared with DSTM [16] approach by employing Polka [15] contention manager. DSTM is the pioneering obstruction-free STM that aborts contentious transactions after consulting with contention manager to ensure progress guarantee for the transactions. Polka contention manager is an accumulation of the best features of Polite and Karma contention management policies. It combines exponential randomized back-off policy of Polite with priority accumulation mechanism of Karma. In [15] it has been shown that Polka gives best overall performance across a wide variety of benchmark for DSTM.

Based on multiple parameters, characterization of transaction is done by randomly generating multiple set of transactions to create data structure micro-benchmarks. Based on these micro-benchmarks, DSTM and AFTM are evaluated. It worth noting that DSTM has not been re-implemented, rather its implementation strategies are executed in the later said benchmarks for a head to head comparison with AFTM.

The performance is measured with respect to multiple set of transactions on the basis of average execution time of Read and Write transactions. In this regard, on the basis of transactions' execution time, set of transactions are segregated into three different groups. The first group contains all the transactions, where average execution time of the Write transactions are less than average execution time of Read transactions (AvgExeTimeWrite < AvgExe-TimeRead). In second group the average execution time of Read and Write transactions are equal to each other with 10% tolerance (AvgExeTimeWrite ≈ AvgExeTimeRead). In the third group, where average execution time of Write transactions are greater than the average execution time of Read transactions are clubbed together (AvgExeTimeWrite > AvgExe TimeRead).

In the experimental process, a set of 15 transactions are being considered at a time. Execution time of transactions in this set varies between 1 and 5 unit-times. This transaction-set is grouped into sub-sets, where each sub-set contains 2–5 transactions. Transactions within a group initiate at the same time. The initiation interval time between each sub-set varies between 1 and 6 unit-times.

Based on these parameters, 5000 set of transactions are generated randomly and segregated into three above said groups. For all these three groups, the λ_i is calculated for every ith transaction, where λ_i = Turnaround Time (TAT)/Execution Time (ET). As TAT = ET + WT, thus, $\lambda_i = 1$ when Wait Time (WT) = 0, otherwise $\lambda_i > 1$. For all the transactions in the transaction set, the κ is calculated as the number of transactions where λ_i of ith transaction is less than λ_{avg}.

4.1 Experimental Results with Increased Number of Write Transactions

The performance comparison between proposed AFTM and DSTM has been done by varying number of write transactions in each set. As the number of write transactions increases, the contention between concurrent transactions is also increased. Thus, transactions' waiting time is higher and hence λavg value is also

Fig. 3 Comparison of λ_{avg}
with varying number of Write
transactions when
AvgExeTime
(Write) < AvgExeTime
(Read)

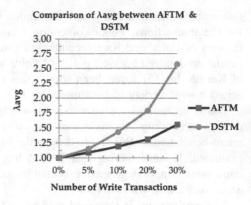

Fig. 4 Comparison of λ_{avg}
with varying number of Write
transactions when
AvgExeTime
(Write) ≈ AvgExeTime
(Read)

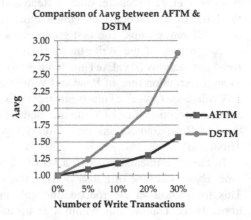

increased. As a result throughput is deteriorated accordingly. Figures 3, 4 and 5 show the experimental results where AFTM performs better than DSTM. At 0% write, when there is no contention between concurrent transactions, WT is 0 and thus λ_{avg} is 1. As the number of Write transactions increases, performance of DSTM deteriorates severely in comparison to AFTTM. When number of Write transactions is increased up to 30%, the λ_{avg} for DSTM is around 60% higher than AFTM. The result set also shows that, whenever average execution time of Write transactions become higher than the Read transactions, performance of DSTM drops down, where as AFTM improves its performance.

In this same scenario, Figs. 6, 7 and 8 depict the performance comparison in terms of κ, i.e. number of transactions where λ_i of ith transaction is less than λ_{avg}. For 5% write transactions, κ is above 90% in case of AFTM, where as in DSTM it is around 87.5% in best case and 75.6% in worst case. When write transactions increases up to 30%, the κ for AFTM is around 70% and for DSTM κ is around 65.05% in their best case scenario.

Fig. 5 Comparison of λ_{avg} with varying number of Write transactions when AvgExeTime (Write) > AvgExeTime (Read)

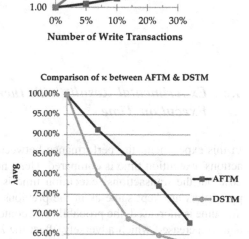

Fig. 6 Comparison of κ between AFTM and DSTM where AvgExeTime (Write) < AvgExeTime (Read)

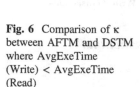

Fig. 7 Comparison of κ between AFTM and DSTM where AvgExeTime (Write) ≈ AvgExeTime (Read)

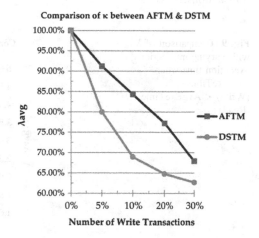

Fig. 8 Comparison of κ
between AFTM and DSTM
where AvgExeTime
(Write) > AvgExeTime
(Read)

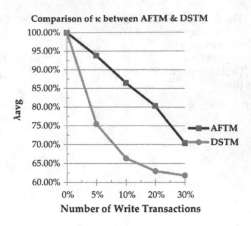

4.2 Experimental Results with Increased Transactions' Execution Time

In this experiment, the performance between AFTM and DSTM based on Transactions' execution time is compared. Here, number of Write transaction is fixed to 20% but the transactions' execution time is being varied. All other experimental parameters are kept same as in the previous experiment. When transaction execution time is increased, the probability of contention between concurrent transactions is also increased which adversely affect the λ_{avg}. Figures 9, 10 and 11 depict these scenarios. The experimental result shows that AFTM outperform DSTM in all three scenarios irrespective of the average execution time of Write and Read transactions. Similar to the previous experiment, AFTM performs better where average execution time of Write transaction is higher than the average execution time of Read transactions.

Fig. 9 Comparison of λ_{avg}
with varying transactions'
execution time where
AvgExeTime
(Write) < AvgExeTime
(Read)

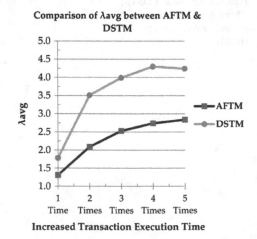

Fig. 10 Comparison of λ_{avg} with varying transactions' execution time where AvgExeTime (Write) $\approx$ AvgExeTime (Read)

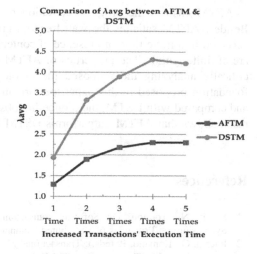

Fig. 11 Comparison of λ_{avg} with varying transactions' execution time where AvgExeTime (Write) > AvgExeTime (Read)

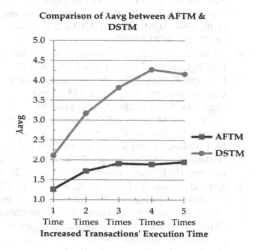

5 Concluding Remarks

In this paper, a new Abort-Free STM (AFTM) has been proposed to handle concurrency between multiple read and write-transactions accessing a shared data item. AFTM does not require aborting any of the contentious transactions. The proposed methodology ensures wait-free, non-blocking implementation where each process completes its operation within a finite number of steps. The proposed AFTM allows multiple transactions to access the same resource concurrently. The operation of a transaction is re-executed in case of data inconsistency. Re-execution of a transaction is better than aborting. This not only avoids the repeated resource allocation overhead, but also avoids using any separate contention manager for ensuring both progress condition and data consistency.

AFTM ensures the progress condition of the system as proved in Lemma 1. Besides, AFTM satisfies the wait-freedom criteria that a group of transactions is executed in a finite time in presence of contention provided individual transactions are of finite length. The properties of AFTM have been proved as Lemmas while critically analyzing the proposed methodology. On the basis of the theoretical foundation proposed in this paper, performance analysis of AFTM has been done and compared with DSTM, the pioneering obstruction free STM. The experimental result shows that AFTM outperforms DSTM in all respect.

References

1. Shavit, N., Touitou, D.: Software transactional memory. In: ACM SIGACT-SIGOPS Symposium on Principles of Distributed Computing, pp. 204–213 (1995)
2. Rachid, G., Romano, P. (eds.): Transactional Memory. Foundations, Algorithms, Tools, and Applications: COST Action Euro-TM IC1001, vol. 8913. Springer (2014)
3. Perelman, D., Fan, R., Keidar, I.: On maintaining multiple versions in STM. In: Proceedings of the 29th ACM SIGACT-SIGOPS Symposium on Principles of Distributed Computing, pp. 16–25. ACM (2010)
4. Attiya, H., Milani, A.: Transactional scheduling for read-dominated workloads. J Parallel Distrib. Comput. **72**(10), 1386–1396 (2012)
5. Cachopo, J., Rito-Silva, A.: Versioned boxes as the basis for memory transactions. Sci. Comput. Program. **63**(2), 172–185 (2006)
6. Fernandes, S.M., Cachopo, J.A.: Lock-free and scalable multi-version software transactional memory. ACM SIGPLAN Not. **46**(8), 179–188 (2011)
7. Riegel, T., Felber, P., Fetzer, C.: A lazy snapshot algorithm with eager validation. In: Distributed Computing, pp. 284–298. Springer, Berlin, Heidelberg (2006)
8. Aydonat, U., Abdelrahman, T.S.: Relaxed concurrency control in software transactional memory. IEEE Trans. Parallel Distrib. Syst. **23**(7), 1312–1325 (2012)
9. Avni, H., Dolev, S., Fatourou, P., Kosmas, E.: Abort free semanticTM by dependency aware scheduling of transactional instructions. In: Networked Systems, pp. 25–40. Springer (2014)
10. Ghosh, A., Chaki, N.: The new OFTM algorithm toward abort-free execution. In: Proceedings of the 9th International Conference on Distributed Computing and Information Technology, pp. 255–266. Springer (2013)
11. Ghosh, A., Chaki, R., Chaki, N.: A new concurrency control mechanism for multi-threaded environment using transactional memory. J. Super Comput. **71**(11), 4095–4115 (2015)
12. Diegues, N., Cachopo, J.: Practical parallel nesting for software transactional memory. In: Distributed Computing, pp. 149–163. Springer (2013)
13. Marathe, V.J., Scott, M.L.: A qualitative survey of modern software transactional memory systems, University of Rochester Computer Science Department, Technical Report (2004)
14. Marathe, V.J., Spear, M.F., Heriot, C., Acharya, A., Eisenstat, D., III, W.N.S., Scott, M.L.: The Rochester software transactional memory runtime. http://www.cs.rochester.edu/research/synchronization/rstm (2015)
15. Scherer III, W.N., Scott, M.L.: Advanced contention management for dynamic software transactional memory. In: Proceedings of 24th annual ACM Symposium on Principles of Distributed Computing, pp. 240–248. ACM (2005)

16. Herlihy, M., Luchangco, V., Moir, M., Scherer III, W.N.: Software transactional memory for dynamic-sized data structures. In: 22nd Annual ACM Symposium on Principles of Distributed Computing, pp. 92–101 (2003)
17. Marathe, V.J., Scherer III, W.N., Scott, M.L.: Adaptive software transactional memory. In: Proceedings of the 19th International Symposium on Distributed Computing (DISC), pp. 354–368 (2005)

Graph Problems Performance Comparison Using Intel Xeon and Intel Xeon-Phi

Jiří Hanzelka, Robert Skopal, Kateřina Slaninová, Jan Martinovič and Jiří Dvorský

Abstract While most modern well known performance benchmarks for high performance computers focused mainly on the speed of arithmetical operations, the increasing amount of nowadays problems depend also on the speed of memory access. This aspect is becoming crucial for all data driven computations. In this paper, two benchmarks focusing on the speed of memory access are examined. The first examined benchmark is well known Graph 500. This benchmark was developed in order to measure the computers performance in memory retrieval using the Breadth First Search algorithm on randomly generated graph. The second benchmark uses the real world data set (Czech Republic traffic network) as an input graph on which the betweenness centrality algorithm is performed. Both of these benchmarks were tested on SALOMON cluster comparing performance on both Xeon processors and Xeon-Phi co-processors. Obtained performance results were analyzed and discussed at the end of the paper.

Keywords Graph 500 · Betweenness centrality · Benchmark · Parallel computing · Xeon · Xeon-Phi

J. Hanzelka (✉) · R. Skopal · K. Slaninová · J. Martinovič · J. Dvorský
IT4Innovations, VŠB – Technical University of Ostrava, 17. listopadu 15, 708 33
Ostrava, Poruba, Czech Republic
e-mail: jiri.hanzelka@vsb.cz

R. Skopal
e-mail: robert.skopal@vsb.cz

K. Slaninová
e-mail: katerina.slaninova@vsb.cz

J. Martinovič
e-mail: jan.martinovic@vsb.cz

J. Dvorský
e-mail: jiri.dvorsky@vsb.cz

© Springer Nature Singapore Pte Ltd. 2017
R. Chaki et al. (eds.), *Advanced Computing and Systems for Security*,
Advances in Intelligent Systems and Computing 567,
DOI 10.1007/978-981-10-3409-1_5

1 Introduction

The significant amount of nowadays computations is becoming data driven. Therefore, a computer ability to fast memory access appears more and more crucial because of the requirements of data processing and analysis [2]. The important area of such computations are big graph problems which are mostly used in order to model large real life networks, i.e. traffic networks, social networks but also in areas like the modeling of connections in the human brain [6].

For example, one of the used algorithms in area of traffic networks is betweenness centrality described in this paper. The general assumption of betweenness centrality in traffic routing is that the most frequently used roads can be considered as bottlenecks [5]. The main goal of betweenness centrality is to identify such bottlenecks in a traffic network represented by oriented weighted graph. This way, betweenness centrality can help to optimize routing since it allows us to monitor how bottlenecks change during the day. Since betweenness centrality is based on the shortest path search, it can also be used to monitor and examine bottlenecks behavior based on wide number of factors. Besides the distance, graph weights could reflect actual weather conditions, speed, time of the day, etc.

Due to these reasons, it becomes highly important to measure and examine the performance of graph algorithms. The authors are the members of the group involved in the Antarex project [1] focused on an autotuning and adaptivity approach for energy efficiency on HPC systems. Betweenness centrality is the algorithm intended to be used within one of the use cases focused on the development of a navigation system. It is very important to investigate and compare different implementations and optimization techniques on the heterogeneous infrastructure of SALOMON cluster for the appropriate usage of this algorithm within the project.

One of the standard widely used graph benchmarks is the Graph 500 benchmark [9] which was created to complement LINPACK benchmarks like well known Top 500 [10]. It emphasizes the speed of memory access instead of the number of floating point operations per second (FLOPS) as in Top 500. The main idea of the benchmark is to perform the Breath First Search (BFS) algorithm on graph generated by Kronecker generator. The detailed description of the Graph 500 benchmark can be found in Sect. 2. For the full specification, see [9].

Due to the fact that Graph 500 works with an artificially generated graph, another benchmark based on betweenness centrality is presented in Sect. 3. Betweenness centrality is a graph metric which is used to identify important nodes in the graph [3]. In this paper, betweenness centrality performance is examined on the graph constructed from the Czech Republic traffic network. Therefore, in this case betweenness centrality will be able to identify important traffic bridges in the network in this case, and this information can be further used for example as an input into routing algorithms.

The performance of both algorithms was tested on SALOMON cluster operated by IT4Innovations National Supercomputing Centre in the Czech Republic using the Xeon processors and Xeon-Phi coprocessors. A detailed experimental setup is described in Sect. 4. The performance is compared side by side using series of experiments is described in Sect. 5. The results were also analyzed using the Intel VTune Amplifier [7]. All the results can be found in Sect. 5.

2 Graph 500 Benchmark

The intent of Graph 500 is to develop a benchmark that has multiple analysis techniques (multiple kernels) accessing a single data structure represented as a weighted, undirected graph. Unlike the well known Top 500 benchmark which measures the number of floating points per second (FLOPS) Graph 500 emphasizes the speed of memory access which is crucial in nowadays data driven computations.

The benchmark considers an undirected weighted graph $G = (V, E)$ where G represents set of vertices and E set of edges. The only input is graph scale S and *edgefactor* which represents number of edges per vertex. Final number of vertices is then $|V| = 2^S$ and number of edges $|E| = edgefactor \times |V|$. Even though there can be arbitrary number of edges in the graph, the default edgefactor requested by the Graph 500 specification is 16 so there are 16 edges per vertex. By knowing this input, the final graph is generated using the Kronecker generator [9].

The overall benchmark performs the following procedure described in the Graph 500 specification [9]:

1. Generate the edge list.
2. Construct a graph from the edge list (timed, kernel 1).
3. Randomly sample 64 unique search keys (vertices) with degree at least one, not counting self-loops.
4. For each search key:

    (a) Compute the parent array (timed, kernel 2).
    (b) Validate that the parent array is a correct BFS search tree for the given search tree.

5. Compute and output performance information.

Note, that only the steps marked as timed are included in the performance information.

In order to measure performance across multiple architectures and programming languages, Graph 500 defines a new metric counting Traversed Edges Per Second (TEPS) [9]. TEPS are measured through the benchmarking of kernel 2 for all 64 keys. The final result is than computed as a harmonic mean of performances from all 64 BFS runs.

Experiments for Graph 500 were done using the Tuned MPI implementation [4] which can be downloaded from the Graph 500 website [9].

3   Betweenness Centrality Benchmark

Since the Graph 500 benchmark works with artificially generated graph, another benchmark based on betweenness centrality using a real graph data set is presented in this chapter.

Betweenness is widely used graph metric used mainly in order to find the most significant vertices within the graph. Betweenness centrality quantifies the number of times a node is a bridge along the shortest path between two other vertices.

The mathematical representation (defined by Freeman in 1977) can be written as

$$C_B(v) = \sum_{s \neq v \neq t \in V} \frac{\sigma_{st}(v)}{\sigma_{st}},$$

where V is a set of vertices, σ_{st} denotes number of shortest paths from $s \in V$ to $t \in V$ (where $\sigma_{st} = 1$ for $s = t$) and $\sigma_{st}(v)$ denotes the number of shortest paths from s to t that $v \in V$ lies on.

For the purpose of the benchmark, the input graph for betweenness centrality is generated from the Czech Republic traffic network map that was obtained from OpenStreetMap project [8]. The traffic network graph is stored in a simple csv file with one edge in each row. The first two columns contain indexes of start and end vertices of each edge and the third column is edge weight. In addition, the first line of the csv file holds total number of vertices in the first column and total number of edges in the second column.

As a performance measurement, the same TEPS metric as described in Sect. 2 was also adapted for the betweenness centrality benchmark.

Betweenness centrality was tested using our C++ implementation of Brande's algorithm [3] that was parallelized by OpenMP and MPI.

4   Experiments

The first experiment was focused on a performance testing of Graph 500 benchmark. We used MPI for parallelization, while for each MPI process we used only one thread. With this implementation, we could use number of MPI processes np according to formula $np = 2n^2$ where n is some positive integer so we choosed such n that the number of processes was near the maximum of available cores on the all allocated nodes.

We ran the tests for different sizes of graph through scale parameter of 1, 5 and 6 nodes. Besides of measuring TEPS value, we used VTune Amplifier 2016 available on SALOMON for the performance testing and we added the most time-consuming methods into the result tables. Values in the tables are from filtered VTune selection from the part responsible for BFS and not for the graph generation.

We recorded the following values for each method (as described in [7]):

ClockTicks number of processor cycles.

CPI Rate measures how many cycles takes each instruction where theoretical best value is 0.25. This metric should be the first one to check in order to optimize the performance.

Front-End Bound this number shows how good is the pipeline filled with useful operations that are fetched and decoded from instructions.

Bad Speculation this metric shows how many operations in the pipeline are actually executed; this value gets higher because of bad instructions prediction.

Back-End Bound this metric shows how good are operations from the pipeline executed; this value gets higher because of data-cache misses.

Retiring This number shows how many executed instructions were actually required.

The higher results indicates where should we focus in order to optimize the performance.

4.1 Experimental Setup

Both Graph 500 and Betweenness were tested on a single node of SALOMON cluster located in IT4Innovations National Supercomputing Centre in Ostrava, the Czech Republic. The SALOMON cluster consists of 1008 computational nodes of which 576 are regular compute nodes and 432 accelerated nodes. The accelerated nodes are accelerated by two Intel Xeon-Phi 7120P cards with 61 cores and 16 GB RAM.

The Code was compiled using the Intel compiler. Only the -xHost flag was added into the original makefile to take an advantage of the Intel AVX2 instruction set and the -mmic flag in order to perform the experiments on the Xeon-Phi accelerator.

In order to use the betweenness centrality algorithm on the traffic network, the initial R-MAT graph generator was replaced with a simple csv parser which is used to load the network from the csv file on the disc and store it into the graph structure in the memory.

The same compiler flags were used as in Graph500.

5 Results

The first experiment was focused on measuring the speed and the performance of the Graph 500 benchmark. The results also contain information about the most time-consuming methods. These methods (with aliases in brackets) were:

- bitmap_matvec_and_mark (Matvec)
- bitmap_matvec_trans_and_mar (MatvecT)

- make-MakeLP (MakeLP)
- dapl_rc_DAPL_recv_dyn_opt_20 (DAPL)

The first three methods are responsible for running BFS algorithm. The method DAPL is responsible for the inter-node communication. Therefore it is why it's in VTune report for 5 and 6 nodes and not only for 1 node.

It can be seen from Tables 1, 2 and 3 can be seen that TEPS are not increasing with larger graph scale. However the results show some spikes. Table 3 contains the results for one node and 18 MPI processes it is visible that there are two good results for scale 22 and 26. There is also visible that the method MakeLP is more important with higher scale and is becoming a bottleneck. In Table 2 for 5 nodes and 50 MPI processes, it is also visible that the DAPL method which is responsible for inter-node communication becomes bottleneck when the small scale is used. On the other hand, the DAPL method is less important with the increased scale, and the MakeLP method becomes the bottleneck again. For 6 nodes and 128 MPI processes the result were the same but TEPS value were higher because more cores were used.

The highlighted values in Tables 1, 2 and 3 show which methods should be optimized according to a particular metric. For example the method MakeLP shows significantly increased values for the CPI rate and Back-End Bound in problem scale 28 in Table 1.. The modern processor should be able to perform four instructions per cycle [7], but the CPI rate of 1.746 indicates that the processor performs approximately only half instruction per cycle. This latency can be caused by cache misses or other bottlenecks. The second highlighted value is for the Back-End Bound metric.

Table 1 VTune report for Graph500 on 6 nodes (144 cores) with 128 MPI processes

Scale	TEPS	Methods	ClockTicks	CPI rate	Front-end	Bad specul.	Back-end	Retiring
24	4.33E+09	MatvecT	8.20E+10	0.659	0.147	**0.428**	0.078	0.348
		Matvec	5.70E+10	0.459	0.105	**0.314**	0.088	0.492
		DAPL	3.20E+10	0.874	0.023	0.072	**0.663**	0.242
		MakeLP	5.00E+09	1.230	0.165	0.059	0.530	0.245
26	4.71E+09	MatvecT	3.21E+11	0.619	0.137	**0.398**	0.100	0.365
		Matvec	2.62E+11	0.510	0.112	**0.312**	0.129	0.448
		DAPL	7.10E+10	**1.161**	0.014	0.011	**0.789**	0.186
		MakeLP	3.40E+10	1.437	0.110	0.040	0.643	0.206
28	4.39E+09	MatvecT	1.21E+12	0.616	0.136	**0.354**	0.138	0.372
		Matvec	1.04E+12	0.528	0.114	**0.283**	0.172	0.431
		DAPL	2.19E+11	**1.060**	0.014	0.034	**0.747**	0.205
		MakeLP	2.31E+11	**1.746**	0.069	0.009	**0.745**	0.177
30	3.51E+09	MatvecT	2.73E+12	0.907	0.125	**0.379**	0.246	0.251
		Matvec	2.22E+12	0.736	0.129	**0.376**	0.185	0.310
		DAPL	3.52E+11	1.155	0.013	0.017	0.774	0.196
		MakeLP	4.32E+11	**1.969**	0.058	0.011	**0.767**	0.164

Table 2 VTune report for Graph500 on 5 nodes (120 cores) with 50 MPI processes

Scale	TEPS	Methods	ClockTicks	CPI rate	Front-end	Bad specul.	Back-end	Retiring
21	1.69E+09	MatvecT	1.23E+10	0.706	0.166	**0.396**	0.098	0.339
		Matvec	7.33E+09	0.424	0.126	**0.308**	0.064	0.502
		DAPL	4.21E+09	0.664	0.020	**0.125**	**0.561**	0.295
		MakeLP	3.70E+08	0.974	0.054	0.014	0.622	0.311
23	1.75E+09	MatvecT	4.90E+10	0.664	0.153	**0.445**	0.058	0.344
		Matvec	3.30E+10	0.449	0.113	**0.292**	0.083	0.512
		DAPL	7.60E+09	0.659	0.013	0.057	**0.594**	0.336
		MakeLP	2.72E+09	1.001	0.103	0.059	0.536	0.303
25	1.97E+09	MatvecT	1.93E+11	0.663	0.153	**0.426**	0.057	0.365
		Matvec	1.42E+11	0.474	0.121	**0.306**	0.083	0.491
		DAPL	2.06E+10	0.662	0.019	0.141	0.548	0.292
		MakeLP	2.72E+09	1.001	0.103	0.059	0.536	0.303
27	1.73E+09	MatvecT	9.08E+10	0.686	0.139	**0.408**	0.118	0.335
		Matvec	6.87E+10	0.519	0.124	**0.32**	0.116	0.440
		DAPL	7.60E+09	0.638	0.026	0.141	0.514	0.320
		MakeLP	1.61E+10	**1.661**	0.047	0.000	**0.754**	0.211

According to [7], this value indicates that one of the possible problems is overloaded divider unit which is responsible for the division of numbers. So in this case, the possible solution would be to reduce the dividing in the code. Another possible source could also be cache misses.

The second experiment was focused on measuring the speed of the betweenness centrality algorithm parallelized using MPI and OpenMP in a hybrid way on a graph generated from the actual Czech Republic traffic network described earlier as an input. Table 4 shows a different number of used threads for one node and also their placement through affinity. We used scatter and compact affinity type. Table 5 shows the results of experiments, in which also the MPI processes were used. In Table 6 are shown hybrid runs with combination of different number of used MPI processes and threads on Xeon Phi (MIC) in native mode to see difference of speed compared to CPU. We used 1 process with 4 threads as minimum for MIC tests because of time limit. Besides measured time, we also added the efficiency and speedup values for the all tests. The base value was the measurement with the lowest used threads (1 for CPU, 4 for MIC).

As seen in Table 4, the difference between scatter and compact affinity is visible for most number of used threads in favor of scatter affinity except when we used 24 threads. In that case, the processor was fully utilized, and the affinity than had no impact. When 24 threads were used, the speedup was 13.88. MPI processes were also used in the next test, as shown in Table 5. It is visible for one node when only MPI processes are used that the efficiency is a little lower than in the case of

Table 3   VTune report for Graph500 on 1 node (24 cores) with 18 MPI processes

Scale	TEPS	Methods	ClockTicks	CPI rate	Front-end	Bad specul.	Back-end	Retiring
21	7.26E+08	MatvecT	6.38E+10	0.687	0.150	**0.453**	0.059	0.337
		Matvec	3.60E+10	0.389	0.093	**0.222**	0.089	0.596
		MakeLP	3.10E+09	1.009	0.114	0.032	0.525	0.328
22	7.58E+08	MatvecT	1.19E+11	0.627	0.114	**0.449**	0.043	0.264
		Matvec	7.50E+10	0.397	0.092	**0.253**	0.081	0.574
		MakeLP	7.70E+09	1.119	0.132	0.020	0.543	0.305
23	7.12E+08	MatvecT	2.47E+11	0.658	0.147	**0.436**	0.071	0.347
		Matvec	1.62E+11	0.432	0.103	**0.260**	0.106	0.531
		MakeLP	2.40E+10	1.394	0.083	0.033	0.644	0.239
24	7.37E+08	MatvecT	8.20E+10	0.671	0.146	**0.417**	0.092	0.345
		Matvec	5.70E+10	0.442	0.110	**0.269**	0.095	0.525
		MakeLP	5.00E+09	**1.534**	0.059	0.012	**0.718**	0.211
25	7.29E+08	MatvecT	1.01E+12	0.660	0.143	**0.405**	0.106	0.346
		Matvec	7.12E+11	0.465	0.115	**0.287**	0.105	0.494
		MakeLP	1.93E+11	**1.676**	0.054	0.024	**0.732**	0.190
26	7.48E+08	MatvecT	1.68E+12	0.625	0.138	**0.387**	0.113	0.362
		Matvec	1.30E+12	0.486	0.119	**0.304**	0.107	0.471
		MakeLP	3.71E+11	**1.816**	0.051	0.027	**0.748**	0.174
27	6.53E+08	MatvecT	2.40E+12	0.706	0.140	**0.414**	0.128	0.319
		Matvec	1.85E+12	0.543	0.128	**0.341**	0.114	0.416
		MakeLP	4.88E+11	**1.905**	0.040	0.024	**0.770**	0.166
28	6.97E+08	MatvecT	5.60E+12	0.644	0.128	**0.332**	0.184	0.355
		Matvec	4.49E+12	0.517	0.115	**0.279**	0.171	0.436
		MakeLP	1.23E+12	**1.986**	0.012	0.027	**0.802**	0.160

OpenMP threads. Using combination of MPI and OpenMP with adding more nodes the speedup increases linearly.

Table 6 contains time, efficiency and speedup of the betweenness centrality algorithm for MIC. It can be observed from the tests that the best efficiency of the betweenness centrality algorithm is when 61 processes and 4 threads per the process is used per MIC. If we use 2 MICs with the same setup then the efficiency is almost the same and adding more MIC increase the speed of the algorithm almost linearly as in the case of adding nodes for the CPU test. For our algorithm, two MICs had a bit slower speed (363 s) in comparison with two CPUs (346 s).

Table 4 Experiments for the betweenness centrality algorithm on 1 node (2 × 12 cores) for different number of used threads

Threads	KMP-affinity	Avg. time [s]	Expected time [s]	Efficiency	Speedup
1	none	4804	4804.00	1.00	1.00
14	scatter	1386	1201.00	0.87	3.47
	compact	1575		0.76	3.05
8	scatter	798	600.50	0.75	6.02
	compact	918		0.65	5.23
12	scatter	582	400.33	0.69	8.25
	compact	676		0.59	7.10
16	scatter	468	300.25	0.64	10.27
	compact	510		0.59	9.42
20	scatter	399	240.20	0.60	12.04
	compact	410		0.59	11.71
24	scatter	346	200.17	0.58	13.88
	compact	346		0.58	13.88

Table 5 Experiments for the betweenness centrality algorithm on 1 node (2 × 12 cores) for different number of used threads and MPI processes

Nodes	MPI processes	Threads	Avg. time [s]	Expected time [s]	Efficiency	Speedup
1	1	1	4804.0	4804.00	1.00	1.00
1	4	1	1773.0	1201.00	0.68	2.71
1	8	1	955.0	600.50	0.62	5.03
1	12	1	665.0	400.33	0.60	7.22
1	16	1	513.0	300.25	0.59	9.37
1	20	1	423.0	240.20	0.57	11.36
1	24	1	362.0	200.17	0.55	13.27
2	48	1	190.6	100.09	0.53	25.20
4	96	1	95.1	50.04	0.53	50.52
8	192	1	47.6	25.02	0.52	100.86
16	384	1	24.0	12.51	0.52	200.17
32	768	1	12.7	6.26	0.49	379.46
32	384	2	12.4	6.26	0.50	386.48

Table 6 Experiments for the betweenness centrality algorithm on MIC (61 cores – 244 HW Threads) for different number of used threads and MPI processes

Used MICs	MPI processes	Threads	Avg. time [s]	Expected time [s]	Efficiency	Speedup
1	1	4	25542	25542.00	1.00	4.00
1	244	1	946	418.72	0.44	108.00
1	61	4	712	418.72	0.59	143.49
2	122	4	363	209.36	0.58	281.45
4	244	4	192	104.68	0.55	532.13
8	488	4	106	52.34	0.50	973.03

6 Conclusion and Future Work

In this work, we tested the performance of Xeon processor and Xeon-Phi coprocessor on large graph problems. We used well known Graph 500 and Betweenness centrality algorithm as benchmarks. The first mentioned benchmark was working with the artificially generated graph and the second benchmark was tested on real data set using the graph generated from the Czech Republic traffic network.

It is obvious from results in Sect. 4 that the performance of one MIC is comparable with performance of one Xeon processor for our implementation of Brande's algorithm [3].

These results are important input for the future work where we will be using betweenness combined with routing algorithms. Due to the large size of the input road graphs its necessary to measure how betweenness behaves on IT4Innovations infrastructure.

In the future we would like to be able to predict betweenness centrality so the values can be exactly computed only several times a day which would save the computational resources and allow us to use approximated values between the computations.

Next future goal will be focusing on the energy efficiency. We would like to measure the power consumption of both processor and MIC and determine how is Xeon-Phi suitable for big graph problems in terms of the energy efficiency.

Acknowledgements This work was supported by The Ministry of Education, Youth and Sports from the National Programme of Sustainability (NPU II) project 'IT4Innovations excellence in science - LQ1602', from the Large Infrastructures for Research, Experimental Development and Innovations project 'IT4Innovations National Supercomputing Center LM2015070' and co-financed by the internal grant agency of VŠB - Technical University of Ostrava, Czech Republic, under the projects no. SP2016/166 'HPC Usage for Analysis of Uncertain Time Series II' and no. SP2016/179 'HPC Usage for Transport Optimisation based on Dynamic Routing II'.

References

1. ANTAREX (AutoTuning and Adaptivity appRoach for Energy efficient eXascale HPC systems), 671623, H2020-FETHPC-2014: http://www.antarex-project.eu/
2. Angel, J.B., Flores, A.M., Heritage, J.S., Wardrip, N.C.: Graph 500 Performance on a Distributed-Memory Cluster. Technical Report HPCF-2012-11 (2012)
3. Brandes, U.: A Faster Algorithm For Betweenness Centrality (2001)
4. Checconi, F., Petrini, F., Willcock, J., Lumsdaine, A., Choudhury, A.R., Sabharwal, Y.: Breaking the speed and scalability barriers for graph exploration on distributed-memory machines. Supercomputing (2012)
5. Fujisawa, K., Matsuo, H.P.: Graph analysis & HPC Techniques for Realizing Urban OS (2015)
6. Lovász, L.: Large Networks And Graph Limits, vol. 60. Colloquium Publications (2012)
7. Intel VTune Amplifier XE: https://software.intel.com/en-us/intel-vtune-amplifier-xe
8. OpenStreetMap: https://www.openstreetmap.org
9. The Graph 500 Benchmark: http://www.graph500.org
10. The Top 500 Benchmark: http://www.top500.org

A Novel Image Steganographic Scheme Using 8×8 Sudoku Puzzle

Debanjali Dey, Ambar Bandyopadhyay, Sunanda Jana,
Arnab Kumar Maji and Rajat Kumar Pal

Abstract The word *steganography* has originated from two Greek words, namely, *stegos* and *grahia* meaning *cover* and *writing*, respectively. The concept behind steganography is that the transmitted message should be indiscernible to the regular eye. It is the art of concealing information within a medium. The main aim of steganography is to protect this concealed information. A message transmitted using a cipher text may incite suspicion, but an invisible message will not. Digital images constitute a flexible medium for carrying secret messages because slight modifications done to a cover image is hard to be distinguished by the human eyes. In this work, we have proposed an Image Steganographic Scheme using 8×8 Sudoku puzzle for secure data transmission. The originality of the work is that if any intruder changes any part of the secret message, then the proposed algorithm will be able to detect it.

Keywords Cover image · Puzzle · Secret message · Steganography · Stego image · Sudoku

D. Dey · A. Bandyopadhyay · R.K. Pal
Department of Computer Science and Engineering, University of Calcutta,
Kolkata 700106, India
e-mail: deydebanjali25@gmail.com

A. Bandyopadhyay
e-mail: ambarbandyopadhyay@gmail.com

R.K. Pal
e-mail: pal.rajatk@gmail.com

S. Jana
Department of Computer Science and Engineering, Haldia Institute of Technology,
Haldia 721657, West Bengal, India
e-mail: sunanda.only1@gmail.com

A.K. Maji (✉)
Department of Information Technology, North Eastern Hill University,
Shillong 793022, Meghalaya, India
e-mail: arnab.maji@gmail.com

© Springer Nature Singapore Pte Ltd. 2017
R. Chaki et al. (eds.), *Advanced Computing and Systems for Security*,
Advances in Intelligent Systems and Computing 567,
DOI 10.1007/978-981-10-3409-1_6

1 Introduction

Nowadays information security is a critical aspect of most communication mediums. Steganography is the art and science of invisible communication; it is the procedure of concealing a file, message, image, or video within another media such as a file, message, image, or video [1]. This is achieved through concealing information in other information. Thus, the existence of the communicated information is hidden. There is a major difference between steganography and cryptography. The latter focuses on maintaining the secrecy of the contents while the former focuses on preserving the existence of a message secret. The steganographic procedure can be described by the following formula:

$$cover\_medium + hidden\_data + stego\_key = stego\_medium$$

In this context, *hidden_data* is the data being hidden/concealed; *cover_medium* is the object in which data is hidden; the data may also be encrypted using a key that is termed the *stego_key*. The resulting file is known as the *stego_medium* that is of the same type of file as the *cover_medium*. The *cover_medium* and *stego_medium* are typically image or audio files. In this work, we have focused on image files and therefore, have referred to them as *cover_image* and *stego_image*.

There are many suitable steganographic techniques, depending on the type of the cover object [2], some of which are listed below:

1. **Image Steganography**: In this technique, an image is taken as the cover medium. Pixel intensities are utilized to conceal the information in this technique.
2. **Network Steganography**: In this technique, network protocol, such as TCP, UDP, ICMP, IP, etc., are taken as the cover object, and the protocol is used as a carrier.
3. **Video Steganography**: In this technique, any file or information is hidden in the digital video format. The video that is a combination of pictures acts as the carrier for the concealed information.
4. **Audio Steganography**: In this technique, audio is taken as a carrier for information concealment.
5. **Text Steganography**: In this technique, the number of tabs, white spaces, capital letters, etc. are used for concealing information.

Further, image steganography techniques can be divided into the following domains:

1. **Spatial Domain Technique**: In this technique, knowledge of the original cover image is required during the decoding process. The decoder checks for differences between the original and the distorted cover images and restores the secret message.
2. **Transform Domain Technique**: In this technique, data is embedded in the frequency domain of a signal, and it is much stronger than time domain techniques.

3. **Distortion Techniques**: There are many variations of spatial steganography, which change some bits in the image pixel values directly, for concealing the data.
4. **Masking and Filtering**: In this technique, information is hidden by marking an image similar to paper watermarks.

In our proposed scheme, image steganography technique has been used to hide the information for embedding secret data in images. An image that is used to cover the secret message is known as the *cover image*. The data that is to be embedded inside the *cover image* is referred to as the *hidden data*. Finally, the key used to encrypt the message is known as the *stego key*. The final output is known as the *stego medium*.

In our proposed work, Sudoku has been used for embedding secret data in images. Sudoku [3] is originally known as Number Place. It is a logic-based, combinatorial number-placement puzzle. The aim is to fill up a 9×9 grid with digits such that each column, row, and the nine 3×3. minigrid that constitute the grid (also called *boxes*, *blocks*, *regions*, or *sub-squares*) contains all the digits from 1 to 9. Completed puzzles are always a type of Latin square with an additional constraint on the contents of individual regions. Apart from 9×9 Sudoku puzzle, other versions of the Sudoku puzzle also exist. They can be briefly classified [3] as follows based on size:

- Sample puzzles with 4×4 grids and 2×2 regions.
- 5×5 grids with pentomino regions have been published under the name *Logi*-5 [4]. A *pentomino* comprises five orthogonally connected congruent squares. *Pentomino* is seen in playing the game Tetris [5].
- The World Puzzle Championship features a 6×6 grid with 2×3 regions [3].
- 7×7 grid with six heptomino [6] regions and a disjoint region.
- 8×8 grid with 4×2 regions.
- *The Times* offers a 12×12 grid 'Dodeka Sudoku' with 12 regions of 4×3 squares [7].
- Dell Magazines regularly publishes 16×16 'Number Place Challenger' puzzles (using the numbers 1 to 16 or the letters A-P) [7].
- Nikoli offers 25×25 Sudoku, the Giant behemoths [7]. A 100×100 grid puzzle dubbed Sudoku-zilla was published in 2010.

This scheme uses an 8×8 Sudoku reference matrix as the key for embedding the secret message into the cover image. Before going to the description about this scheme, we have to know about how an image is represented. An image [8] is an array, or a matrix, of square pixels organised in rows and columns. In an 8-bit grayscale image, each picture element is allocated an intensity ranging from 0 to 255. A greyscale image is a term normally assigned to a black and white image, but the name emphasises that it also includes several shades of grey. A normal grayscale image has 8-bit colour depth i.e. 256 greyscales. A 'true colour' image has 24-bit colour depth i.e. $8 \times 8 \times 8$ bits $= 256 \times 256 \times 256$ colours ~ 16 million colours. This scheme uses a 24-bit cover image for concealing the secret message.

In 24-bit RGB schemes, a single pixel is represented using 3 bytes, and each byte signifies 8 bits of either red, blue, or green component information.

A standard Sudoku is a logic-based, combinatorial number-placement puzzle consisting of 9×9 grids. In this work, we have defined a Sudoku Matrix as an 8×8 matrix with the numbers from 0 to 7, such that each number occurs exactly once in each row, exactly once in each column, and exactly once in each region.

Figure 1 illustrates an instance of 8×8 Sudoku puzzle and its solution has been given in Fig. 2. We have termed the solution of the Sudoku puzzle a Sudoku matrix.

Fig. 1 An instance of 8×8 Sudoku puzzle

1		4			7	6	
	7	0	6			5	
4	1	6		7	3	0	2
2		3	7	5		1	4
0	3			6	4	2	5
5		2	4	0	1		
7	2	5		1		4	6
6				2		3	7

Fig. 2 A solution of 8×8 Sudoku puzzle referred as Sudoku matrix

1	5	4	2	3	7	6	0
3	7	0	6	4	2	5	1
4	1	6	5	7	3	0	2
2	0	3	7	5	6	1	4
0	3	7	1	6	4	2	5
5	6	2	4	0	1	7	3
7	2	5	3	1	0	4	6
6	4	1	0	2	5	3	7

2 Existing Literature

The first work on steganography using Sudoku puzzle was by Chang et al. [9] published in 2008. This paper proposed a new approach on steganography using Sudoku puzzle with a digital signature for providing authentication, integrity, as well as non-repudiation.

In 2010, a maze-based steganographic system was proposed by Lee et al. [10]. In this proposed scheme, to hide the secret data, a maze game was used as the carrier media. This proposed scheme considered multi-path rather than the lone solution path to gain more embedding capacity [10].

In 2012, Saradha et al. [11] proposed a method for improving data hiding capacity using Sudoku puzzles in colour images. The main idea of the method was to use a Sudoku puzzle where each value relates to a pixel pair (red, blue) of the image embedded with the secret data by replacing a pair of one pixel of two colours. This method was proposed to improve the visual quality of the stego image and also improve the average hiding capacity of the image to 4 bits/pixel. This scheme makes use of 24 bits of any image and modifies 16 bits of each pixel. In this technique, the Sudoku solution is utilized for both embedding the secret data into the image and extracting it from the image.

Tawade et al. proposed an efficient data hiding scheme [12] using secret reference matrices where data was hidden in the 8-bit grayscale image using a 256×256 matrix. This matrix was constructed using a 4×4 table with unreported digits from 0 to 15. The proposed method improves the holding capacity of cover image and increases the complexity to crack the Secret Reference Matrix. A new data hiding scheme using spatial domain was also proposed by them using a secret reference matrix for data embedding and extraction.

In 2014, Job and Paul [13] introduced an image steganography technique using Sudoku puzzle and ECC algorithm to enhance the security of data transmission. In this method, image steganography hid the data into a picture, while the ECC algorithm helped to convert the original data into the secret code. In the proposed method, the image steganography was done using Sudoku puzzle. Thus, this method not only hid the data, but it also converted the original data into a secret code by combining these two techniques.

Recently in 2015, Usha et al. [14] proposed an enhancement upon the work of Ijeri et al. [15] wherein a hard Sudoku was used instead of a soft one. In the first step, the embedding and extraction algorithms were generalized to work with any $K \times K$ sudoku where K is a perfect square. Then a comparison between the hard and soft Sudoku techniques was made. Moreover, a study was done for the different Sudoku sizes to compare parameters such as PSNR and payload capacity.

However, the main drawback of these schemes is that most of the schemes are applied to greyscale images. Moreover, they have used tile matrix of size 9×9. In this paper, we have used a steganographic scheme with 8×8 Sudoku puzzle which is of lesser dimension. Moreover, the scheme can be applied to colour images.

3 Proposed Technique

All the existing techniques have used 'TILE Matrix'. To make this 'TILE Matrix' extra storage space is required. Another drawback of these methods is that they use Sudoku matrix of size 9×9 or 16×16 or 25×25 that takes too much space. Another drawback is the data embedding capacity. We have seen that a reference matrix of varying sizes has been used. The range lies between 256×256 to 18×18. However, use of a smaller reference matrix reduces the computation time. Still, it is a time-consuming task which is yet another drawback.

From this perspective, we have used an 8×8 reference matrix that is much smaller in size. We have seen that it can embed a secret message into a gray-scale image. The input is a colour image that is first converted to a gray-scale image. Then the secret message has been embedded by manipulating pixel values and finally the resultant gray-scale image has been converted to a colour image. It is clear that it is time-consuming, and that is another drawback. If any intruder changes the pixel values, then it is not possible to report the corrupted image which is the most dangerous drawback among all. Now in this section, we have proposed a new algorithm to overcome the drawbacks of the existing methods. The proposed scheme has been shown in Fig. 3.

Here, we have used an 8×8 Sudoku matrix as shown in Fig. 2 that has been used to encrypt and decrypt the secret message. We have used the term 'secret message' for the data which is going to be embedded and 'cover image' for that image into which the message is going to be embedded. To encrypt the secret message into the cover image, we have first converted the ASCII value of each

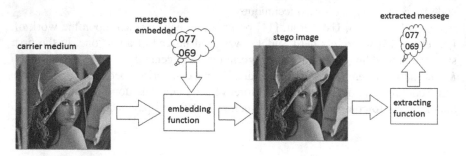

Fig. 3 The proposed scheme

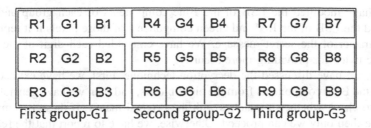

First group-G1   Second group-G2   Third group-G3

Fig. 4 The structure of the Sudoku cell

character of the secret message to the base-8 number system and then have stored the base-8 value of each character in an array. We have assumed that the row and column of the matrix start from 0.

For example, the secret message is 'H'. The ASCII value of 'H' is 72 and after converting 72 to the base-8 number system, we got 110. We have to embed 110 into the cover image. We have calculated the length of the secret message and then divided it by 64 to know how many blocks are needed for this reason. As we have used an 8 × 8 matrix, and each cell of this matrix represents one character of the secret message, thus, 1 block represents 64 characters. The maximum length of the secret message of our proposed algorithm is proportional to the height and width of the cover image.

We have grouped 9 pixels into one cell. The 9 pixels from one cell have formed three subgroups $G1$, $G2$, and $G3$. It is evident from the above diagram that each cell contains 3 columns, the first column contains 3 pixels which we have referred to as $G1$, second column as $G2$ and third column as $G3$. The structure has been shown in Fig. 4.

At first, we have converted the value of the cell into binary number system. As the range of the value lies between 0 to 7, we need only 3 bits to represent this value. LSB of Blue components of each group will represent this binary value.

Returning to our example, 'H' has been embedded into (0, 0) cell of the Sudoku matrix. The value of the cell (0, 0) is 1. The first digit of the secret message is 1 so our target is to find the position of 1 in the first row and first column as 1 is the value of the cell. It is clear that the value of the cell will behave like an index. We have found that 1 has been placed in (1, 7) and (2, 1) cells. We have embedded this 7 into the LSB of the Red components of each pixel belonging to group $G1$, and 2 into the LSB of the Green components of each pixel belonging to group $G1$. The second digit of each character of the secret message have been embedded into $G2$ and the third digit into $G3$.

We have modified the LSB of the Blue components of each group and have modified the LSB of the Red and Green components of those pixels if it represents any character of the secret message. We have embedded each digit of the secret message twice for providing more security.

Next, we have discussed the decryption technique. First, we have extracted the LSB of the Blue components from the eight blocks, and then we have generated the matrix to check whether it matches with the original matrix. If these two matrices are same then only we can proceed. Otherwise, we have to report that the received image has been corrupted.

If the image has not been corrupted, then we can decrypt the secret message. From group $G1$ of the cell, we have to extract the value of LSB of Red and Green components. Suppose the value of the cell is x and the value obtained from the Red and Green components are y and z, respectively. If the value of the cells (x,y) and (z,x) are the same, then we have to store the value and do the same for $G2$ and $G3$.

After getting one character containing three bits of the secret message we have converted the value to the decimal number system. This value denotes the ASCII value of the character and then we have stored the character. Continuing this way, we have decrypted the remaining characters from the required blocks following the procedure mentioned above.

Returning to our example, the LSB of the Red and Green components of $G1$ represent 7 and 2, respectively. The LSB of the Red and Green components of $G2$ represent 7 and 2, respectively. The LSB of the Red and Green components of $G3$ represent 2 and 3, respectively. It can be seen that the value of the cells $(1, 7) = (2, 1)$ and the value is 1. The value of the cells $(1, 2) = (3, 1)$ and the value is 0. Combining these we have obtained $(110)_8$. Converting it into the decimal number system, we have obtained $(72)_{10}$ and it represents 'H' which is the secret message. Therefore, we can conclude that we have successfully decrypted the secret message. Here, we have use '$' as the delimiter to indicate the end of the secret message.

3.1   Algorithm for Encryption

```
Input   : Cover image, Sudoku matrix, and Secret message.
Output  : Cover image in which secret message is embedded.

Step 1  : Start.
Step 2  : Read the all pixel value of the cover image and
          the secret message.
Step 3  : Calculate the length of the secret message.
Step 4  : If the length of the secret message is greater
          than the embedding capacity, then print an error
          message and exit.
Step 5  : Else convert ASCII value of each character into
          base-8 number system.
Step 6  : If the length of any character after converting
          into Base-8 number system is not 3, then pad 0
          from left accordingly.
Step 7  : For i = 0 to 7 do
Step 8  : For j = 0 to 7 do
Step 9  : Read a character from the secret message.
Step 10 : Set x = matrix (i,j)
Step 11 : LSB of Blue component from each group represents
          the value of x.
Step 12 : LSB of Red component from G1 belonging to the
          cell represents column number of x^{th} row in which
          the MSB of the scanned character is placed.
Step 13 : LSB of Green component from G1 belonging to the
          cell represents row number of x^{th} column in
          which the MSB of the scanned character is
          placed.
Step 14 : Repeat step 12 and 13 for the group G2 and G3 to
          embed second digit and third digit of the
          scanned character.
Step 15 : If any character left, then only follow Steps 7
          through 14.
Step 16 : Else skip Steps 12, 13, and 14.
End for of Step 8
End for of Step 7
Step 17 : Stop.
```

3.2   Algorithm for Decryption

```
Input   :  Cover image in which secret message is embedded.
Output  :  Secret message.

Step 1  :  Start.
Step 2  :  Extract LSB of Blue component from each group.
Step 3  :  Make a matrix with these extracted 64 values
           and then compare it with the original matrix.
Step 4  :  If it matches, then go to Step 2 to do the same
           procedure for the remaining part of the stego-
           image.
Step 5  :  If these and the original matrix are the same,
           then
           Step 5.1  :  For i = 0 to 7 do
           Step 5.2  :  For j = 0 to 7 do
           Step 5.3  :  Set x = matrix(i,j)
           Step 5.4  :  Extract LSB of Red component from
                        each pixel belonging to the group G1
                        of a block and read it as y.
           Step 5.5  :  Extract LSB of Green component from
                        each pixel belonging to the group G1
                        of a block and read it as z.
           Step 5.6  :  If matrix(x,y) = matrix(z,x), then
                        store the value and go to step 5.4
                        for groups G2 and G3.
           Step 5.7  :  else go to step 6.
           Step 5.8  :  Combined the extracted three values
                        in that order G1, G2 and G3 and then
                        convert it into a decimal number
                        system and store the character.
           End for of Step 5.1
           End for of Step 5.2
Step 6  :  Else report for corrupted image.
Step 7  :  Stop.
```

3.3   Advantages of This Proposed Technique

- High data embedding capacity (proportional to the height and width of the cover image).
- We have ensured that if any intruder changes the secret data, then it will be detected successfully at the receiver end.

- We have seen that all the existing techniques at first convert the colour image to a grayscale image and then encryption is done. After that, the greyscale image is again converted to a colour image. Therefore, it is time-consuming. However, here we have directly manipulated the RGB values. Thus, the proposed algorithm takes less time from the perspective of execution time.
- A Minimal amount of space is needed to store 8×8 matrixes.
- Here, we have not used any TILE matrix. Therefore, from this point of view, the proposed technique also reduces storage space.
- Now, it becomes unnecessary to send the Sudoku matrix explicitly as it has been embedded into the cover image.
- The distortion of any component can be maximum 1 here.

4 Experimental Results

The cover image distortion varies with the changes in the pixel values. The number of pixels of the cover image used for embedding depends on the number of components of the pixel used and the amount of input data. We have used *PSNR* to evaluate the quality of an image. The *PSNR* is defined as follows:

$$PSNR = 10\log_{10}\left(\frac{255^2}{MSE}\right)\text{dB},$$

where *MSE* is the mean square error between the original image and the stego image. The *MSE* is defined as follows:

$$MSE = \frac{1}{3} * V * W\left(\sum_{i=0}^{V-1}\sum_{j=0}^{W-1}\left\{\left(R_{ij} - R_{ij}^l\right)^2 + \left(G_{ij} - G_{ij}^l\right)^2 + \left(B_{ij} - B_{ij}^l\right)^2\right\}\right)$$

$V * W$ gives the number of all pixels present in an image. A larger *PSNR* indicates that the quality of the stego image is closer to the original one. Normally human's eyes find it hard to distinguish between the distortions on a stego image compared to original image when its *PSNR* value is greater than 30 dB.

Here, we have used the Fig. 5 for the experiment whose width and height is 512×512 and throughout the discussion, we have used the image shown in Fig. 5.

4.1 Comparison of Experimental Results

The PSNR value of existing methods are in greyscale so to compare these with our proposed method we take the average of PSNR value of RED, GREEN, and BLUE

Fig. 5 An image used for the experimentation

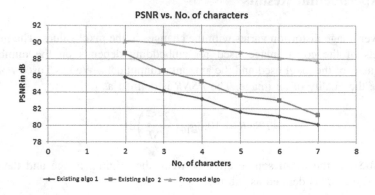

Fig. 6 Graphical comparison between existing and proposed method

component of pixels and plot it in an X-Y plane (see Fig. 6). We plot a number of characters along the X-axis and PSNR (dB) along the Y-axis.

4.2 Analysis of Experimental Results

Tables 1 and 2 illustrate the results of the proposed technique and the comparison of the proposed and existing methods, respectively. By analysing the results of Table 2, it can be clearly seen that the *PSNR* values of the proposed technique are very high compared to the *PSNR* values of the existing techniques. The results clearly indicate that the *PSNR* values of the proposed method are robust. The *PSNR* values have been taken for 21 pure-text messages and 3 numbers with characters for the existing and the proposed methods. Thus, it can be concluded that the proposed method performs better than the existing methods.

Table 1 Performance results (*PSNR*) of proposed technique

Message	PSNR (dB)		
	Red	Green	Blue
So	95.3604985	91.5583861	84.4262816
Me	93.3192987	91.2107650	84.4262816
My	95.3604985	90.8889182	84.4262816
Us	95.3604985	90.8889182	84.4262816
We	93.8992181	91.9362717	84.4262816
Bye	94.5686860	90.8889182	84.4262816
You	96.3295986	90.5892859	84.4262816
Six	93.3192987	90.3089987	84.4262816
Out	93.8992181	90.3089987	84.4262816
Law	94.5686860	91.2107650	84.4262816
Find	92.3501985	90.5892859	84.4262816
Hang	92.3501985	90.5892859	84.4262816
Lock	91.9362717	90.8889182	84.4262816
Okay	92.8077734	90.8889182	84.4262816
Goal	91.9362717	90.8889182	84.4262816
Hello	91.2107650	89.5626625	84.4262816
Array	92.3501985	90.3089987	84.4262816
Basic	92.3501985	90.5892859	84.4262816
Point	92.8077734	89.7974735	84.4262816
Turbo	92.8077734	90.3089987	84.4262816
%56#	91.5583861	88.0365609	84.4262816
^&*123	91.5583861	89.5626620	84.4262816
#58974*	90.0457093	88.7329202	84.4262816
Calcutta University	85.9156718	85.0262609	84.4262816

Table 2 Average PSNR in dB for *n* characters

No. of characters	Existing algorithm		Proposed algorithm		
	Grey level	Grey level	Red	Green	Blue
2	85.81132	88.63408	94.66000	91.29665	84.42628
3	84.17294	86.58102	94.53710	90.66139	84.42628
4	83.18240	85.27170	92.27614	90.76907	84.42628
5	81.62497	83.59476	92.18085	89.76733	84.42628
6	81.09470	82.97630	90.80205	89.14779	84.42628
7	80.09170	81.23310	90.04571	88.73292	84.42628

Figure 7 shows the histogram of the original and the stego image, and here we have embedded a secret image whose length is 511 characters. These two images cannot be distinguished visually because the resultant stego-image is less distorted. Therefore, it can be concluded that the proposed technique is efficient.

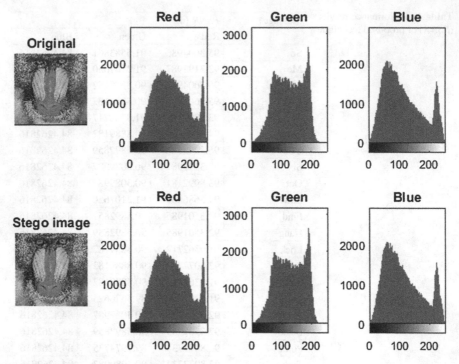

Fig. 7 Histogram analysis of images

If any stego image is too much distorted, then anyone can guess that something has been added to the original, and then anyone can intentionally change the pixel values. However, here both the histogram is almost equal, and therefore intruder may think that the stego-image is not the stego-image, rather it is the original image.

We have also performed steganalysis of the *stego_image* and we have found that the proposed scheme in this paper is safe against geometrical attack, noise based attack, etc.

4.3 Complexity Measurement

In this section, we discuss the complexity issue. The proposed method is compared with two existing methods in the literature [16]. They have embedded the cipher image into the cover image, and for that, a Sudoku matrix is used with the help of genetic algorithm (GA). The authors have not discussed any computational complexity of their method and even the earlier existing method, to which they compared their results, was also a GA based article [17]. Theoretically, that too takes an infinite amount of time and space. Hence, in this section, we only discuss the complexity issue of our proposed method.

In our proposed method, to embed a single character of the secret message, we need nine pixels (Fig. 4). Now, each secret character contains three digits, and each digit is embedded into each group. Thus, to embed each digit we have to fetch two LSBs (Least Significant Bits) of R (Red) and G (Green) part of each pixel. Thus, the number of computations needed to embed a single character is equal to 6 × 3 or 18. As a result, if the secret message contains m characters, then the total number of computations needed is 18 × m.

Now, we are also embedding the value of the Sudoku cell into LSB (Least Significant Bit) of B (Blue) part of each pixel of the cover image. Then, the maximum number of computations needed to embed the value into that position is L × W, if the dimension of the image is L × W. Thus, the total number of computations needed is (18 × m + L × W). Hence, the computational complexity of the proposed method is $O(n^2)$ in the worst case.

5 Conclusion

Data transmission is an important thing what everyone needs because everyone wants his or her data to be transmitted securely. This paper has proposed a secure methodology based on a steganographic technique using a Sudoku matrix. The Sudoku matrix has been used for embedding the cypher into the cover image. Thus, the data can be transferred more securely by embedding the cypher into an image. Hence, whenever data need to be transmitted securely, this technique can be implemented.

In this work, we have proposed a steganographic technique, where we have used an 8 × 8 Sudoku reference matrix for message embedding. We have also embedded that 8 × 8 Sudoku matrix into the cover image for checking whether the cover image has been modified or not. This 8 × 8 matrix is the key component of the proposed algorithm. If an intruder changes any LSB value of any pixel, then the cover image gets modified, and it can easily be detected as we have already embedded the 8 × 8 Sudoku matrix inside it. It can also prevent any modification as each Sudoku puzzle matrix should have values 0 through 7 only once in the same row, in the same column, and in a minigrid. An 8 × 8 Sudoku reference matrix has been used for hiding the secret message, whereas the earlier existing methods used 256 × 256 or 27 × 27 or 18 × 18 reference matrix. That is why less computation is involved in our method. Therefore, it can be concluded that our proposed scheme is more robust with less computation.

References

1. https://en.wikipedia.org/wiki/Steganography
2. Hussain, M., Hussain, M.: A survey of image steganography techniques. Int. J. Adv. Sci. Technol. **54**, 113–124 (2013)
3. http://en.wikipedia.org/wiki/Sudoku
4. http://www.en.wikipedia.org/wiki/Pentomino
5. http://www.en.wikipedia.org/wiki/Tetris
6. http://mathworld.wolfram.com/Heptomino.html
7. http://en.wikipedia.org/wiki/Glossary_of_Sudoku
8. https://www.spacetelescope.org/static/projects/fits_liberator/image_processing.pdf
9. Chang, C.C., Chou, Y.C., Kieu, T.D.: An information hiding scheme using Sudoku. In: Proceedings of Third International Conference on Innovative Computing, Information and Control (ICICIC 2008) (2008)
10. Lee, H.L., Lee, C.F., Chen, L.H.: A perfect maze based steganographic method. J. Syst. Softw. **83**(12), 2528–2535 (2010)
11. Saradha, P., Swamy, B.: Improving image data hiding capacity scheme using Sudoku puzzle in colour images. Int. J. Eng. Res. Appl. **2**(3), 2741–2744 (2012)
12. Tawade, L., Mahajan, R., Kulthe, C.: Efficient & secure data hiding using secret reference matrix. Int. J. Netw. Secur. Appl. **4**(1), 43–50 (2012)
13. Job, D., Paul, V.: Image steganography technique using Sudoku puzzle and ECC algorithm for secured data transmission. J. Theoret. Appl. Inf. Technol. **66**(2), 447–459 (2014)
14. Usha, B.A., Srinath, N.K., Anvith, E.: High capacity data embedding in image steganography using hard Sudoku puzzle. Int. J. Adv. Res. Comput. Commun. Eng. **4**(6), 518–525 (2015
15. Sanmitra, I., Pujeri, S., Shrikant, B., Usha, B.A.: Image steganography using Sudoku puzzle for secured data transmission. Int. J. Comput. Appl. **48**(17), 31–35 (2012)
16. Chithra, B., Harinath, D., Murthy, M.V.R., Babu, K.R.: Enhancing security through steganography by using Sudoku puzzle and ECC algorithm. Int. J. Res. Sci. Eng. Technol. **2** (6), 29–43 (2015)
17. Rajakumar, B.R.: Static and adaptive mutation techniques for Genetic algorithm: A systematic comparative analysis. Int. J. Comput. Sci. Eng. **8**(2), 180–193 (2013)
18. Gupta, S., Goyal, A., Bhushan, B.: Information hiding using least significant bit steganography and cryptography. Int. J. Modern Educ. Comput. Sci. **6**, 27–34 (2012)

Part II
Data Analytics

Comparison of K-means Clustering Initialization Approaches with Brute-Force Initialization

Martin Golasowski, Jan Martinovič and Kateřina Slaninová

Abstract Data clustering is a basic data mining discipline that has been in center of interest of many research groups. This paper describes the formulation of the basic NP-hard optimization problem in data clustering which is approximated by many heuristic methods. The famous k-means clustering algorithm and its initialization is of a particular interest in this paper. A summary of the k-means variants and various initialization strategies is presented. Many initialization heuristics tend to search only through a fraction of the initial centroid space. The final clustering result is usually compared only to some other heuristic strategy. In this paper we compare the result to the solution provided by a brute-force experiment. Many instances of the k-means can be executed in parallel on the high performance computing infrastructure, which makes brute-force search for the best initial centroids possible. Solutions obtained by exact solvers [2, 11] of the clustering problem are used for verification of the brute-force approach. We present progress of the function optimization during the experiment for several benchmark data sets, including sparse document-term matrices.

1 Introduction

Clustering is a basic exploratory data analysis technique from a family of unsupervised machine learning methods. Its purpose is to identify and form clusters of similar data based on a distance metric between individual data points. Beside the distance metric it requires no additional knowledge about statistical properties of the data set.

M. Golasowski (✉) · J. Martinovič · K. Slaninová
IT4Innovations, VŠB – Technical University of Ostrava,
17. Listopadu 15/2172, 708 33 Poruba, Ostrava, Czech Republic
e-mail: martin.golasowski@vsb.cz

J. Martinovič
e-mail: jan.martinovic@vsb.cz

K. Slaninová
e-mail: katerina.slaninova@vsb.cz

© Springer Nature Singapore Pte Ltd. 2017
R. Chaki et al. (eds.), *Advanced Computing and Systems for Security*,
Advances in Intelligent Systems and Computing 567,
DOI 10.1007/978-981-10-3409-1_7

Clustering represents a powerful tool in many fields [17], including computer vision, chemistry, biology, financial markets, etc.

Its main objective is to form both *homogenous* and *well-separated* clusters. *Homogenity* of a cluster is achieved by grouping similar data points together as close as possible. *Separation* of clusters means that distance between individual clusters should be maximized. Many surveys have been performed on clustering techniques that satisfy this condition [17]. It has been shown that good clustering (in terms of homogeneity and separation) can be achieved through minimization of squared-sum distances of all the points to centers of their respective clusters.

Minimal sum of squared distances clustering (MSSC) can be formally described as follows. Partition given data set $X = \{x_1, x_2, ..., x_n\}$ of n d-dimensional vectors to k-partitions $C = \{c_1, c_2, ..., c_k\}$ such that the sum of squared distances of the individual vectors to centers of their respective clusters is minimized through the entire partitioned data set. Each cluster has its own center m_j determined by mean of the data points assigned to that cluster. The MSSC can be described as a following optimization problem:

$$\underset{w,m}{\text{minimize}} \sum_{i=1}^{n} \sum_{j=1}^{k} w_{ij}(x_i - m_j)^2$$

$$\text{subject to} \sum_{j=1}^{k} w_{ij} = 1, \forall i = 1, ..., n \tag{1}$$

$$w_{ij} \in \{0, 1\}, \forall i = 1, ..., n; \forall j = 1, ..., k$$

where w_{ij} is the association weight of a given data point x_i to cluster c_j. Cluster center is determined as mean of its associated data points as follows:

$$m_j = \frac{\sum_{i=1}^{n} w_{ij} \cdot x_i}{|C_j|} \tag{2}$$

where $|C_j|$ is number of data points associated with cluster j.

It has been shown that the MSSC problem is NP-hard [1] for $k <= 2$ in general dimension. However, large number of various heuristics for approximating the optimal solution of the MSSC has been widely known for quite some time. Namely, one of the most common approaches known as the k-means algorithm has been subject of intensive research efforts for several decades [17]. Although the idea of k-means was first conceived by Hugo Steihaus in 1956 [24], the standard algorithm was proposed by Stuart Lloyd in 1957 (although published later in 1982) [20]. The same method was published earlier by Forgy in 1965 [13]. Other widely known variants are the MacQueen's algorithm [21] and the Hartigan-Wong algorithm [15].

1.1 Variants of the Clustering Problem

The objective function of the general clustering problem presented above measures sum of the squared Euclidean distances which is perfectly suitable for problems in the Euclidean space (such as image processing). The data are dense in this type of problems, often with smaller number of dimensions. However, the main purpose of the data clustering is to observe patterns in large and multi-dimensional datasets which are impossible to visualize using standard means. One such type of data are text documents which are usually represented as long sparse vectors of relative term frequencies. Euclidean distance in this case is impractical due to the well-known curse of dimensionality. The clustering problem can be generalized [10] as

$$\underset{w,m}{\text{minimize}} \sum_{i=1}^{n} \sum_{j=1}^{k} w_{ij} d(x_i, m_j) \tag{3}$$

where $d(x_i, m_j)$ is the used distance measure.

The *cosine dissimilarity* measure can be used to overcome the curse of dimensionality and is determined by

$$d(\mathbf{x}, \mathbf{m}) = 1 - \cos(\mathbf{x}, \mathbf{m}) = 1 - \frac{\mathbf{x} \cdot \mathbf{m}}{|\mathbf{x}| \, |\mathbf{m}|} \tag{4}$$

where $\mathbf{x}, \mathbf{m}$ are the two input vectors of the same dimension. This measure computes an angle between the vectors while suppressing the influence of their actual length. Values of this measure are within the range $\langle 0; 1 \rangle$, where 0 means orthogonality of the two vectors and 1 means their correspondence. K-means variant described by Dillon in [10] is called *spherical k-means* and uses this distance measure. In this paper, we present experiments performed with implementation available in the R statistical computing environment [22] in package sk-means described in [16].

1.2 Initialization

Sensitivity to the selection of initial centroids is one of the main disadvantages of the algorithm. Quality of the results obtained from slightly different set of initial centroids may vary greatly. It is also quite easy to find instances of the problem for which the algorithm will not converge in time.

A significant amount of research has been put into overcoming this hurdle and many different approaches has been developed [17, 25]. The simplest methods randomly select *k* points from the data set [13] or select *k* first points as the centroids [21]. According to recent extensive survey presented in [8], those methods behave rather poorly as the selection process does not look at the statistical properties of the whole data set. One of the more complex method is the kmeans++ method [3] which randomly selects only the first initial centroids and rest of the centroids are selected with probability determined by:

$$P(x) = \frac{d_{min}(x)^2}{\sum_{j=1}^{N} d_{min}(x_j)^2} \tag{5}$$

where d_{min} is a minimum distance from the current point x to the previously selected centers. The intention of this algorithm is to maximize distance of the selected centroids. Similar method has been proposed by Bradley and Fayyad [6] which tries to identify areas with the highest density, thus with bigger concentration of good candidate points. The method first creates J subsamples of the data set, runs a standard k-means initialized by selecting first k points as centroids from the subsample. Resulting centroids for the J runs of the k-means are combined together and clustered by k-means one more time. Centroids obtained from the last run of the k-means are then used as initial centroids for the main clustering run.

The survey of initialization methods [8] shows that performance of the last two mentioned methods is significantly better than the rest of the tested methods (i.e. random initialization). Since the space of the initial centroids (selected as points from the data set) is finite, we would like to explore behavior of the k-means algorithm initialized, if possible, by running all of the possible combinations of the initial centroids or at least running a large sample of the centroids to observe behavior of the criterion function by using high performance computing approach. The observation can then be used as a baseline for the development of the initialization heuristics.

This paper is structured in the following sections. Section 2 discusses the differences between the exact solution of the clustering problem and the heuristics. A brief summary of the exact methods is presented. Methods used to help the heuristics to obtain a better local optimum of the clustering function are also briefly mentioned. Section 3 discusses a proposed brute-force approach. Section 4 contains results of the brute-force search of the initial centroid space for both euclidean and spherical k-means variants.

2 Baseline

Two surveys published by Steinley in 2006 [25] and by Jain in 2010 [17] provide a representative summary of a vast research effort that has been put into improving clustering heuristics. Many different extensions and amendments to the original k-means have been proposed. Many proposals justify correctness and improvement of their methods by an experimental verification. Such benchmarking is often performed either by using synthetic or well-known real world data. Sometimes both internal and external metrics of validity are used as minimization criteria depending on availability of the ground truth data, however shown improvement is only relative to some other (older) method while global optimum of the minimized function (1) is neither quantified nor mentioned [9, 19].

2.1 Global Optimum

As k-means is a heuristic algorithm that offers an approximate solution the clustering problem, it is capable of finding only the local minimum of the optimized function (1). In 2003, Likas et al. published a new heuristic called global k-means that focuses on overcoming this disadvantage of standard k-means [18]. Their method incrementally runs k-means for $k \in \{k - 1, k - 2, \ldots, k - (k - 1)\}$. The initial centroids are obtained by combining the $k - 1$ results from the previous iteration and adding a point from the data set. In each iteration the k-means is executed for a set of n initial centroids and the best results with the lowest value of the Eq. 1 are used as an input in the next iteration. This approach can be impractical for large data sets since $n \cdot (k - 1)$ instances of k-means has to be executed. However, the authors have been able to obtain near optimal results for standard data sets by comparing the obtained results with the exact methods which will be mentioned later in Sect. 2.2.

This method gained a significant amount of attention and several groups of authors proposed various verifications and extensions of the method. Hansen et al. [14] provide an interesting evaluation of the basic global algorithm. They have shown that even for small dimensions there are instances of the clustering problem for which the global algorithm is unable to find an optimal solution. Bagirov et al. [4, 5] provide an extension of the global algorithm which focuses on refining the searched space and elimination of unnecessary k-means runs. Their method optimizes the memory usage of the algorithm and provides better run times when compared to the original proposal.

2.2 Exact Methods

One way of finding global minimum of the MSSC problem is to determine all possible partitions of a given set and to compute sum of squared error for each partition. The number of all possible partitions of a set can be determined by computing its Bell number [26]. To demonstrate size of the problem, let's compute this number for a well known Iris data set [12] which is widely used as benchmark data and as an example of a real world clustering problem. The Iris data set has 4 attributes in 150 data points which represent measured lengths of leaves of three Iris plants types. The Bell number for this data set is approximately 6.82×10^{192}.

Based on the fact that the Iris data set contains exactly three groups of data (for three different kinds of plant) we can determine a number of possible partitionings of the data to the three groups. This can be computed as Stirling number of the second kind [26] and for the three groups of the Iris data set the number is approximately 6.17×10^{70}. The computational infeasibility of brute-force solution is obvious for relatively small data set even without rigorous mathematical proof [1].

These facts do not lessen the need for setting a ground truth baseline for benchmarking of the clustering algorithms. Several exact methods for solving the MSSC problem have been published in recent years. In 1999 du Merle et al. published a paper in which optimal clustering of the Iris data set has been presented for the first time. Their approach is based on various methods of mathematical programming, including column generation, interior point methods, hyperbolic programming, a variable neighborhood search heuristic, and quadratic 01 programming [7, 11]. They have been able to compute exact solution to a number of standard benchmark data sets but only for a small number of instances (approx. 200). This method was later improved by Aloise [2] by devising a new geometric approach to the auxiliary problem. The authors of the improved methods have been able to solve much larger data sets (approx 2000). Brusco et al. [7] proposed a different method based on a repetitive branch and bound procedure. They have been able to solve smaller data sets (approx. 200 instances) but for larger values of $k \geq 9$. In [2], the authors present a comparison of the mentioned methods for a set of standard benchmark data sets.

3 Brute Force Initialization

As explained earlier in Sect. 1.2, the k-means sensitivity to initialization centroids can be avoided by using an appropriate initialization method. Deterministic methods often tries to select initial centroids in high density regions of the data set and as far from each other as possible. Another methods are based on random sampling of s initial centroid vectors and subsequent initialization of the k-means. The best initial centroids can be picked by selecting the clustering result which optimizes the clustering criterion (3) or selected cluster validation index (Random Index, Dunn, etc.).

These methods share a common goal which is to shrink the space of candidate initial centroids by eliminating the candidates with unsatisfying properties (i.e. centroids too close to each other, outliers, noise, etc.). However, as many of those methods are heuristics, they may find only approximate best initial centroids. Searching the entire space of the candidate initial centroids means running k-means for $\binom{n}{k}$ times for each possible combinations of the initial centroids (assuming we pick initial centroids as one of the existing data points from the data set).

In this paper, we propose an experiment in which we determine if some of the euclidean k-means initialization methods offers satisfying results when compared to the best possible result obtained by searching through the entire space of possible initial centroids. The baseline for the euclidean k-means obtained in this experiment is then compared to the results given by the exact MSSC methods. The brute-force solution is naturally infeasible for a larger data set, however, by running sufficiently large number of randomly initialized k-means instances, the behavior of the criterion function can be observed.

4 Experiments

The main purpose of the experiments is to observe whether initialization heuristics are able to provide similar results to those obtained by exhaustive search (brute-force) solution. We present results of the standard k-means algorithm on three small datasets in the first part of the experiment. The next part of the experiment tries to use brute force initialization on a large multi-dimensional dataset using spherical k-means.

We expect that the convergence on the small datasets will be fast which would allow us to compare results of various initialization methods. However, due to their relatively small size and well-defined structure, the datasets do not reflect real world situations. The experiment with the large dataset using spherical k-means can provide hint on behavior of the rate of the criterion function (Eq. 4), even though exhaustive search of the initial centroids space cannot be performed in reasonable time.

The experiments were executed on three data sets. The first one widely used— Ruspini data set [23] contains $n = 75$ data points in 2 dimensions, having 4 well-separated spherical clusters. The second one is a synthetic data set containing $n = 150$ data points in 2 dimensions, having 2 distinct clusters. Both of the data sets are unlabeled. The third data set is the well known Fisher's Iris [12] containing $n = 150$ data points with 4 attributes.

Another experiment tests how many random initializations have to be executed for k-means to converge on a large unstructured dataset. This experiment has been performed on the well-known Medlars data set containing text of 1033 medical abstracts. The data has been pre-processed to the document-term matrix, where rows represents the individual documents and columns relative frequencies of 8707 terms. Large number of random initializations can be executed by using HPC cluster just to observe the convergence of the clustering criterion function. Graph representation of the Medlars data set generated by Gephi software based on cosine similarity matrix after removing the weakest ties is presented in Fig. 1.

The experiments were implemented using the R statistical computing environment [22]. Created implementation was parallelized using tools from the *parallel* and *Rmpi* packages and executed on nodes of the Anselm cluster managed by the IT4Innovations National Supercomputing Center. The cluster consists of 209 compute nodes, each having 64 GB of RAM and two 8-core Intel Xeon® (Sandy Bridge) CPUs.

4.1 *Euclidean K-Means*

In the following experiment, we compare results obtained by three different initialization methods with referential values obtained by exact methods mentioned in Sect. 2.2. The first method f_{random} randomly samples 1000 sets of k initial centroids with uniform probability.

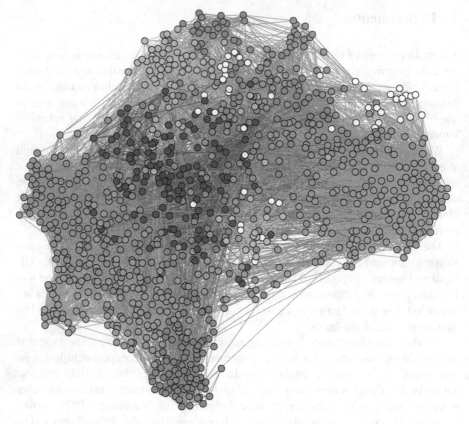

Fig. 1 Graph representation of the Medlars data set, cosine similarity

The second method f_{comb} generates every possible k combination of n data points which results in set of all possible initial centroids. This method is unfeasible for practical use due to its computational demands but can be applied as a benchmark reference for sufficiently sized data set and for small values of k given enough CPU power (e.g. Iris data set).

The third method f_{global} is the unamended version of *global k-means* introduced in [18]. The main difference from the other (brute force) methods is that it heuristically tries to minimize the sum of squared error during its run while searching through all possible initializations in each step. Although this method does not perform an exhaustive search of the initial centroid candidate space, it provides very good results in reasonable time, and it is also more practical for real world usage due to its complexity $\mathcal{O}(kn)$ than the other two methods. This method is described in Sect. 2.1.

A comparison of the results is presented in Table 1. There are minimal values of the clustering criterion for each data set and multiple values of k. The values of k have been chosen with complexity of the f_{comb} method. The last column of the table (f_{ref}) contains MSSE values obtained for a given data set and k by the exact methods

Table 1 The best obtained values of the clustering criterion for different datasets and initializations

Data set	k	f_{random}	f_{comb}	f_{global}	f_{ref}
Iris	2	152.4	152.4	152.4	152.4
	3	78.9	78.9	78.9	78.9
	4	57.2	57.2	57.2	57.2
Ruspini	2	89 337.8	89 337.8	89 337.8	89 337.8
	3	51 063.4	51 063.4	51 063.4	51 063.4
	4	12 881.0	12 881.0	12 881.0	12 881.0
*Synth	2	805 252.4	805 252.4	805 252.4	–
	3	562 241.4	562 241.4	562 241.4	–
	4	444 982.6	444 982.6	444 982.6	–

presented in [2]. Exact values are available only for the first two data sets; the synthetic data set has been generated by our own procedure, thus the exact solution has not been determined yet.

The result of the comparison is evident; each tested method has been able to provide result equal to the global minimum obtained by the exact methods. Based on this experiment we can assume that various methods developed to lessen the k-means sensitivity to bad initialization are indeed able to find such initial centroids that lead to a best possible result.

4.2 Spherical K-Means

Minimal values obtained by randomly initialized spherical k-means are presented in Fig. 2. The graph shows minimal values obtained for $k = 10$ by running n random trials and selecting the best (lowest) obtained value of the spherical k-means objective function (3) from each run. Each random initialization was performed by selecting k random points from the dataset. For each value $n - i$ runs of the k-means has to be performed where $i \in \langle 1; n \rangle$. In total $\sum_{i=1}^{n} i$ runs of the spherical k-means has been performed. The best value obtained from the all runs is 659.66, while the worst value was 666.95.

We have determined that approx. 100 random initializations of the algorithm for given values of k are needed to achieve the convergence, since 100 random initializations yield best value of 660.27 which is very close to the best value obtained after 500 initializations. A fairly good clustering based on the value of the target function can be obtained by performing sufficient number of random initializations even on large unstructured multi-dimensional dataset.

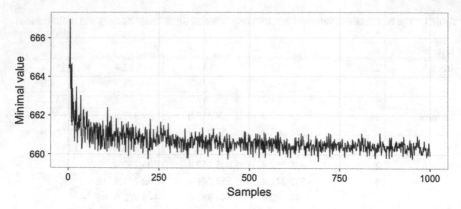

Fig. 2 The best value of the criterion function by a number of random samples for $k = 10$

Table 2 Run times of the initialization methods (in seconds)

Data set	k	t_{random}	t_{comb}	t_{global}
Iris	2	1.388	10.147	0.569
	3	0.978	463.015	0.540
	4	0.964	14 025.600	0.679
Ruspini	2	0.513	1.016	0.262
	3	0.417	25.335	0.207
	4	0.478	456.415	0.207
Synth	2	1.366	10.690	0.601
	3	0.970	475.622	0.594
	4	0.905	14 892.14	0.779
Medlars	10	883.823	–	–

4.3 Run Times

Table 2 contains runtimes of the individual methods for different values of k and the three experimental data sets. Runtimes of the the global k-means method are the smallest since this method searches through the smallest space—$n \cdot k$ candidate centroids. In the case of the combination methods, the size of the input space is obvious even for a small k where the time spent for $k = 3$ is almost 50 times longer than for $k = 2$. All the experiments with the standard euclidean k-means were performed on two 8 core CPUs of a compute node of the Anselm cluster. The experiment with the spherical k-means on the Medlars data set was performed using 64 compute nodes of the Anselm cluster.

5   Conclusion and Future Work

This article provides a summary of the recent research focused on the k-means clustering heuristics. One of the main points of this article is to show that a significant gap exists between the experimental (practical) and the theoretical k-means research. Many publications focus on formalizing of its mathematical properties (NP-hardness [1], exact methods [2, 7, 11]). Simultaneously, there are many contributions that propose various enhancements of the k-means without taking into account amount of theoretical research that has been put into k-means over the years. This insufficiency is often evident in parts where actual improvement of the new enhanced method is presented. While there are methods which can provide exact results for some of the standard benchmark data sets, the authors often present only relative improvement by comparison to some previous variant of the algorithm without (sufficiently) referring to the actual optimization problem that is approximated by the k-means [9, 19].

In this article, we present the exact clustering result [2] for two benchmark data sets and compare them to the results obtained by the k-means initialized using the three different methods. We have found that some heuristics can provide the same results as the exact methods in comparable times. We have also found that heuristics focused on searching of the global best solution [18] can provide same results as the brute force methods using the exhaustive search of the input space of the initial centroids.

However, we cannot experimentally prove that this is valid for larger data, since the exact clustering solutions have been published only for a handful of standard data sets. We have presented results obtained by running a large number of random initializations on the unstructured multi-dimensional data set. We have achieved a good convergence in sufficient time even though a small number of the initial centroids input space for the large text dataset has been searched. This way, by increasing the number of random initializations a good clustering result can be obtained without the need to use a complex initialization heuristics.

Exact methods presented in [2] have significant computational requirements. The best solution has been obtained only for relatively small data sets. Similar results can be possibly obtained for larger data sets by creating their highly optimized implementation and running it in parallel on a high performance computing cluster. We would like to explore a possibility to use a massive parallel accelerators such as GPUs or Intel MIC architecture.

Acknowledgements This work was supported by The Ministry of Education, Youth and Sports from the National Programme of Sustainability (NPU II) project 'IT4Innovations excellence in science—LQ1602' and co-financed by the internal grant agency of VŠB—Technical University of Ostrava, Czech Republic, under the projects no. SP2016/179 'HPC Usage for Transport Optimisation based on Dynamic Routing II'.

References

1. Aloise, D., Deshpande, A., Hansen, P., Popat, P.: Np-hardness of euclidean sum-of-squares clustering. Mach. Learn. **75**(2), 245–248 (2009)
2. Aloise, D., Hansen, P., Liberti, L.: An improved column generation algorithm for minimum sum-of-squares clustering. Math. Program. **131**(1–2), 195–220 (2012)
3. Arthur, D., Vassilvitskii, S.: k-means++: the advantages of careful seeding. In: Proceedings of the Eighteenth Annual ACM-SIAM Symposium on Discrete Algorithms, pp. 1027–1035. Society for Industrial and Applied Mathematics (2007)
4. Bagirov, A.M.: Modified global k-means algorithm for minimum sum-of-squares clustering problems. Pattern Recogn. **41**(10), 3192–3199 (2008)
5. Bagirov, A.M., Ugon, J., Webb, D.: Fast modified global k-means algorithm for incremental cluster construction. Pattern Recogn. **44**(4), 866–876 (2011)
6. Bradley, P.S., Fayyad, U.M.: Refining initial points for k-means clustering. In: ICML. vol. 98, pp. 91–99. Citeseer (1998)
7. Brusco, M.J.: A repetitive branch-and-bound procedure for minimum within-cluster sums of squares partitioning. Psychometrika **71**(2), 347–363 (2006)
8. Celebi, M.E., Kingravi, H.A., Vela, P.A.: A comparative study of efficient initialization methods for the k-means clustering algorithm. Expert Syst. Appl. **40**(1), 200–210 (2013)
9. Deng, G., Tao, J., Zhou, M., Xu, Y.: Improved k-means algorithm with better clustering centers based on density and variance. In: Industrial, Mechanical and Manufacturing Science: Proceedings of the 2014 International Conference on Industrial, Mechanical and Manufacturing Science (ICIMMS 2014), 12–13 June 2014, Tianjin, China. vol. 1, p. 147. CRC Press (2015)
10. Dhillon, I.S., Modha, D.S.: Concept decompositions for large sparse text data using clustering. Mach. Learn. **42**(1–2), 143–175 (2001)
11. Du Merle, O., Hansen, P., Jaumard, B., Mladenovic, N.: An interior point algorithm for minimum sum-of-squares clustering. SIAM J. Sci. Comput. **21**(4), 1485–1505 (1999)
12. Fisher, R.A.: The use of multiple measurements in taxonomic problems. Ann. Eugen. **7**(2), 179–188 (1936)
13. Forgy, E.W.: Cluster analysis of multivariate data: efficiency versus interpretability of classifications. Biometrics **21**, 768–769 (1965)
14. Hansen, P., Ngai, E., Cheung, B.K., Mladenovic, N.: Analysis of global k-means, an incremental heuristic for minimum sum-of-squares clustering. J. Classif. **22**(2), 287–310 (2005)
15. Hartigan, J.A., Wong, M.A.: Algorithm as 136: a k-means clustering algorithm. J. Roy. Stat. Soc. Ser. C (Appl. Stat.) **28**(1), 100–108 (1979)
16. Hornik, K., Feinerer, I., Kober, M., Buchta, C.: Spherical k-means clustering. J. Stat. Softw. **50**(10), 1–22 (2012)
17. Jain, A.K.: Data clustering: 50 years beyond k-means. Pattern Recogn. Lett. **31**(8), 651–666 (2010)
18. Likas, A., Vlassis, N., Verbeek, J.J.: The global k-means clustering algorithm. Pattern Recogn. **36**(2), 451–461 (2003)
19. Liu, H., Fang, C., Wu, Y., Xu, K., Dai, T.: Improved k-means algorithm with the pretreatment of PCA dimension reduction. Int. J. Hybrid Inf. Technol. **8**(6), 195–204 (2015)
20. Lloyd, S.: Least squares quantization in PCM. IEEE Trans. Inf. Theory **28**(2), 129–137 (1982)
21. MacQueen, J., et al.: Some methods for classification and analysis of multivariate observations. In: Proceedings of the Fifth Berkeley Symposium on Mathematical Statistics and Probability. vol. 1, pp. 281–297, Oakland, CA, USA (1967)
22. R Core Team R: A Language and Environment for Statistical Computing. R Foundation for Statistical Computing, Vienna, Austria (2015). http://www.R-project.org/
23. Ruspini, E.H.: Numerical methods for fuzzy clustering. Inf. Sci. **2**(3), 319–350 (1970)
24. Steinhaus, H.: Sur la division des corp materiels en parties. Bull. Acad. Polon. Sci **1**(804), 801 (1956)
25. Steinley, D.: K-means clustering: a half-century synthesis. Br. J. Math. Stat. Psychol. **59**(1), 1–34 (2006)
26. West, L.J., Hankin, R.K.: Set partitions in r. J. Stat. Softw. **23**(02) (2007)

Association Based Multi-attribute Analysis to Construct Materialized View

Santanu Roy, Bibekananda Shit and Soumya Sen

Abstract Analysis of data is an inherent part in the world of business to identify interesting patterns underlying in the data set. The size of the data is usually huge in the modern day application. Searching the data from the huge data set with a lesser time complexity is always a subject of interest. These data are mostly stored in tables based on relational model. Data are fetched from these tables using SQL queries. Query response time is an important quality factor for this type of system. Materialized view formation is the most common way of enhancing the query execution speed across industries. Different approaches have been applied over the time to generate materialized views. However few attempts have been made to construct materialized views with the help of Association based mining algorithms and none of those existing Association based methods measure the performance of the views in terms of both Hit-Miss ratio and view size scalability. This paper proposes an algorithm which generates a materialized view by considering the frequencies of the multiple attributes at a time taken from a database with the help of Apriori algorithm. Apriori algorithm is used to generate frequent attribute sets which are further considered for materialization. Moreover by varying the support count, changing the sizes of the frequent attributes sets; proposed methodology supports scalabilisalubrityty as well as flexibility. Experimental results are given to prove the enhanced results over existing inter-attribute analysis based materialized view formation.

Keywords Association rules · Materialized view · Apriori algorithm · Frequent attribute set

S. Roy
Department of MCA, Future Institute of Engineering & Management, Kolkata, India
e-mail: santanuroy84@gmail.com

B. Shit
Department of CSE, Future Institute of Engineering & Management, Kolkata, India
e-mail: bibek.blues@gmail.com

S. Sen (✉)
A.K. Choudhury School of Information Technology, University of Kolkata, Kolkata, India
e-mail: iamsoumyasen@gmail.com

© Springer Nature Singapore Pte Ltd. 2017
R. Chaki et al. (eds.), *Advanced Computing and Systems for Security*,
Advances in Intelligent Systems and Computing 567,
DOI 10.1007/978-981-10-3409-1_8

115

1 Introduction

Business Analytics and Intelligence have become major area of interest for organizations over the last few decades. In order to survive and excel in the competitive
market scenario, organizations are becoming more and more inclined to use
information technology enabled business intelligence tools. Traditionally these
tools are deployed on a data warehouse over historical data. The volume of this data
is huge. In large databases and data warehousing applications, query response time
plays an important role as timely access of information is the basic requirement of
this type of computations. Different techniques exist for faster query execution such
as materialized view construction, index formation, etc. Among these materialized
view is very much flexible because it could be constructed based on preferred
attributes, user specified size of data and in a manner independent of applications.
Moreover materialized view behaves exactly like tables. This motivates the perspective researchers and organizations to construct materialized view. The difference between a normal view and materialized view is that normal view occupies no
space (other than that for its definition in the data dictionary). A materialized view
occupies space. It exists in the same way as a table: it sits on a disk and could be
indexed or partitioned. Materialized views ensure the availability of frequently
accessed data so that the query execution is optimized. As the data inside a
materialized view is summarized data set instead of large tables, the availability of
desired data inside a view becomes very critical in the performance of the materialized view. Any query generated from the user end, first searches the materialized
views instead of the large database. However, if the desired data is not present in the
materialized views then the original tables are accessed to get the result of the
query. Availability of desired data in the materialized views is termed as hit and the
non-availability of desired data is termed as miss. A better hit-ratio is an indication
of well performing materialized view. Many techniques have evolved over the
period to construct materialized views. There have been numerous research works
in the field of materialized view formation and maintenance. In the next section a
literature survey is made on the existing works on materialized views.

2 Related Work

Many techniques have evolved over the period to construct materialized views. In
this section we focus on some of the useful techniques to form materialized views.
Materialized view formation is often employed in database system as well as data
warehouse and OLAP systems. The parameters like view complexity, query access
frequency, execution time and update frequency of the base table to select a subset
of views from a large set of views to be materialized were considered in dynamic
cost model [1]. Another work [2] figures out the relevant dynamic materialized
views for a given query and an algorithm is proposed to find a small set of relevant

views that can answer a query. Exact and appropriate functional dependencies and Conditional Functional Dependencies (CFD) re redefined over previous summarization techniques such as condensed and quotient cubes [3] to form materialized views. In another approach, materialization of cuboids is experimented on QC-tree which is most storage efficient structure for data cubes in MOLAP. Though high compression ratio is achieved by QC-tree, still it is a fully materialized data cube. An algorithm was proposed in [4] to select and materialize some of the cells instead of the traditional methods requiring all the cells. Modern day database system majorly run on relational databases and the corresponding OLAP application is called ROLAP. In relational model data is represented in the form of table consisting of attributes and tuple. Measurement of attribute affinity gives an idea about the relationship among attributes. In [5] a numeric scale is constructed to enumerate the strength of associations between independent data attributes. However [5] don't propose any methodology to construct materialized view. In another research work linear regression [6] method is used to measure the inter association among attributes and this knowledge is used to form materialized view. The drawback of [5, 6] is that both compute attribute relationship as a pair (two attributes) at a time to construct materialized views.

A clustering technique based method is proposed in [7]. The drawback of [7] is that it never tests the association between more than two attributes together. Though in [7] the validity of materialized view formed as a cluster is tested where it is checked whether two or more attributes can be the members in a single cluster but still that is checked by pair wise attribute relationship.

An improvement over [6] is proposed in [8] by considering relationship between attributes using Non-linear regression method. According to [8] scientific or physical processes/data are in general inherently nonlinear. So Linear Regression performs poorly to identify relationships among attributes. The authors in [8] have theoretically and experimentally shown improved performance over the methodology discussed in [6] based on Hit-Miss ratio.

Using data clustering techniques [9], view materialization for data mining was proposed to generate effective results. If the data are continuously processed and are streamed data then also materialized view can be formed in a dynamic way. Further, different research has been going on to extract the optimal data set to be used for materialization. Earlier research work, as described in [10] had shown that the optimal materialized view selection was an NP-complete problem and the same research work had also proposed a greedy algorithmic based approach for view materialization to optimize query evaluation cost. The approach shown in [10] was dependent on a data structure called data cube. As the first commercial database package, Oracle databases have used the materialized view with a large volume of data and this was discussed in [11]. Different research papers have done comparative studies on different approaches for view selection. One such review study was done in [12] and it was shown that a greedy algorithmic based approach with a polynomial time complexity would have been an optimal way for view selection for materialization. Based on the greedy algorithmic approach, a cost model was developed in [13]. In that work, different calculations were made on evaluation of the total cost and the benefits involved in each materialized view selection and based on the outcome, the most optimized materialized view was selected for a data warehouse. Data Mining techniques have been adopted in [14] to select candidate materialized views and indexes in a data warehousing environment. The authors in [14] have proposed a methodology to couple materialized views and indexes together to achieve efficient storage cost. A comprehensive survey of the different approaches and algorithms to construct and maintain materialized views is presented in [15]. According to the survey work the existing methodologies for selection of materialized views have been divided into three major approaches. (a) Data Mining based Approaches (20%), (b) Heuristic Approaches (35%) and (c) Randomized Algorithmic Approaches (45%). Each algorithms belonging to these three approaches have been analyzed in terms of time and space complexity [15]. Finally the major challenges and issues of these existing methodologies have been discussed in [15]. A randomized view selection algorithm, SSVA (Simulated Annealing View Selection Algorithm) based on simulated annealing to select Top-K views among all possible sets in a multi dimensional lattice, is presented in [16]. Experimentally it has been shown in [16] that SSVA performs better than established greedy algorithm HURA [10] in terms of TVEC (Total View Evaluation Cost). An improvement over the algorithm proposed in [16] is made by the same set of authors in the research work presented in [17]. A randomized view selection two phase optimization algorithm VS2POA is presented in [17]. In the first phase the algorithm selects the best locally optimized Top-T views. These Top-T views become the initial set for the second phase which is based on simulated annealing. Experimental results prove the better performing views in higher dimension.

3 Few Concepts Needed for Proposed Algorithm

In this section some of the concepts and terms which are needed by the proposed methodology are discussed.

3.1 Attribute Usage Matrix

An m × n binary valued matrix in which the value of each entry denoted by use (A_j, Q_i) is either 0 or 1. Let $Q = \{Q_1, Q_2, ..., Q_n\}$ be the set of user queries (applications) that will run on relation $R(A_1, A_2, ..., A_n)$.

Then, for each query Q_i and attribute A_j, Attribute Usage Value, denoted by use (A_j, Q_i) is defined as use $(A_j, Q_i) = 1$ if attribute A_j is referenced by query Q_i, else use $(A_j, Q_i) = 0$.

3.2　Attribute Set

Attribute Set is considered as analogous to Itemset in transactional commercial database. It signifies the set of attributes which are accessed together by a particular transaction or query. In this research work items are treated as attributes. So from here on we would refer itemset as Attribute Set.

3.3　Minimum Support Count

Minimum Support is the user defined threshold value which is minimum count of queries or transactions that should be supported by any Attribute Set at any level.

3.4　Frequent Attribute Set

An attribute set which satisfy minimum support count. For example, if $\{A_1, A_2, A_3, A_4, A_5, A_6\}$ is the relation schema minimum support count is taken as 2, if only $\{A_1\}$, $\{A_2\}$, $\{A_5\}$ and $\{A_6\}$ satisfy the minimum support count then we say $\{A_1\}$, $\{A_2\}$, $\{A_5\}$ and $\{A_6\}$ are frequent 1-attribute set, similarly on satisfying the minimum support count $\{A_1, A_2\}$, $\{A_5, A_6\}$ is 2-attribute set.

3.5　k-Attribute Set

k is an integer (1, 2, ..., n). For example, $\{A_1\}$ is a 1-attribute set, $\{A_1, A_2\}$, $\{A_2, A_3\}$ are 2-attribute set, $\{A_1, A_2, A_3\}$ is a 3-attribute set.

3.6　Query Frequency Matrix

Numerous Queries are generated by the users which access the different attributes of the Attribute Usage Matrix. Any entry in the Query Frequency Matrix corresponding to any query Q_i indicates total frequency of query Q_i.

4　Motivating Factors to Identify the Research Gap

In Sect. 2 we discussed about the three techniques [15] of forming materialized views. According to [15] the Data Mining based approaches are the least explored with only 20% of the existing works. In Data Mining, Association Rule mining and clustering are the two most significant techniques to explore new patterns and identify similar objects. The study of the existing works on Data Mining based approaches confirms that clustering based approaches have overwhelmingly dominated the Association based approaches. Majority of the previous research works (Clustering based approaches as well as other Non-Data Mining approaches) have considered relationship or association between attributes as pairs (two attributes at a time). The drawback of most of the works is that the association between more than two attributes together is not considered. For example, to calculate relationship between attributes A_1, A_2 and A_3 together, the previous methods would calculate association between (A_1, A_2) and (A_2, A_3) or (A_1, A_3) and (A_2, A_3). This approach transitively produces the measure of relationship among A_1, A_2 and A_3 together. The drawback of this approach is that pair wise calculation on attributes increases the computational complexity. Moreover due to the transitive calculation the precision of the computation may be compromised. Therefore it is preferable to have the capability to calculate the relationship among more than two attributes at a time. This research gap is identified in this work. Association based algorithms have the capability to do multi-attribute analysis at a single instance of time. Association based algorithms work on a transactional database to find frequent itemsets. In the context of database involving attributes instead of items sold in the commercial market, each transaction can be taken as the execution of a query where each query accesses a set of attributes. Very few works have been carried out on materialized views using Association based approach. Almost all the methodologies of [14, 18, 19] using Association rules, are actually a combination of clustering and frequent itemset mining algorithms. Moreover none of the existing Association based research works have shown any comparative experimental result analysis in terms of two key quality measurement factors of a materialized view, i.e. Query Hit-Miss Ratio and View size scalability. There are few popular association based algorithms such as Apriori Algorithm [20], FP Growth Algorithm [21] and DIC Algorithm [22]. Among them Apriori is the most frequently used association based algorithm for identifying the frequent item sets from large transactional database. Apriori is most efficient during the candidate generation process. It uses a breadth first search strategy to count the support of item sets and uses a candidate generation function which exploits the downward closure property of support. Apriori uses pruning techniques to avoid measuring certain itemsets, while guaranteeing completeness. Thus the proposed research work has considered Apriori algorithm for finding association between the multiple attributes and identifying the frequent attribute sets for the formation of materialized views. In this work materialized view is formed by the proposed methodology on two separate Attribute Usage Matrices. After that a comparative analysis is made between the existing Non-Association based methods and the proposed algorithm shows superior performance.

5 Use of Apriori Algorithm in the Proposed Methodology

In the proposed methodology at first we have used Apriori algorithm to identify the frequent attribute set(s) and then this knowledge is used to form the materialized views.

5.1 Relevance of Apriori Algorithm in the Construction of Materialized View

Apriori algorithm is used for frequent itemset mining and association rule learning. Since we have related itemset with Attribute Set in this work, finding out Frequent Attribute Set using Apriori algorithm is important in this context. From these frequent attribute sets the proposed algorithm finds out combinations of attributes which are strongly associated to form the views. At any level k, there could be k number of attributes which are closely associated. These k numbers of attributes are known as k-Attribute Set (Discussed in Sect. 2.2). Similarly at level (k + 1), Apriori algorithm gives a measure of the association of (k + 1) number of attributes. Therefore at different levels, Apriori algorithm shows the association among different numbers of attributes. Hence the major contribution of this research work is the formation of materialized views by analyzing more than two attributes at a time, whereas, the existing research works with only two attributes at a time. Once the frequent attribute sets are identified, they are considered for materialization.

5.2 Overview of Apriori Algorithm

The algorithm uses a level-wise search, where k-Attribute Sets are used to explore (k + 1)-attribute sets, to mine frequent attribute sets from transactional database. In this algorithm, frequent subsets are extended one attribute at a time and this step is known as candidate generation process. Then groups of candidates are tested against the data. The algorithm terminates when no further frequent attribute sets are generated at a new level.

5.3 Apriori Property

The Apriori principle states, if an attribute set is frequent, then all of its subsets must also be frequent. Algorithm proposed in this research work is based on the above principle.

6 Materialized View Formation Using Proposed Algorithm

In this section we describe the proposed methodology in Sect. 5.1. In Sect. 5.2 the proposed methodology is presented in the form of an algorithm.

6.1 Proposed Methodology to Construct Materialized Views

The proposed algorithm to form materialized views in this research work has two parts. First, it generates Frequent Attribute Set (Item Set) from the candidate attribute sets at each level using the Apriori algorithm. The constructed Attribute sets at level k are considered to be materialized views of size k. For example in Sect. 2.4 if the frequent attribute set at level 2 is $\{A_1, A_2\}$, then it is considered to be a materialized view of size 2. If during the generation of materialized view the size of view is declared as k, then k number of attributes would form the materialized view. Therefore the proposed algorithm would be executed until it generates the k- frequent attribute sets. For example if a large relation contains 10 attributes and only 40% of the attributes are desired inside a view then the algorithm would be used to generate 4 frequent attribute sets. There may be two cases in this scenario. (i) The algorithm may fail to generate any frequent attribute sets at kth level and stop at some earlier level and (ii) The algorithm may generate frequent attribute sets at levels higher than k. In case (i) the attribute sets formed at the final level would be returned and considered for materialization. For example if the algorithm stops at level 3 forming 3-frequent attribute sets instead of 4-attribute sets. In this situation the algorithm considers 3-frequent attribute sets as the final attribute sets and infers that 4 attribute set generation is not possible. In case (ii) the algorithm will stop it's execution at level 4 returning the 4-frequent attribute sets for materialization. The higher levels would be discarded.

As input the algorithm needs a relation, the set of user queries executed on that relation, the minimum support count and the size of the materialized view (Say k). At first it generates the m × n Attribute Usage Matrix (AUM) from the relation R and set of queries applied over R (where m is the number of queries and n is the number of attributes). The AUM is taken as the input for the Apriori algorithm to generate k frequent attribute sets. The difference between the Apriori algorithm and the traditional algorithm is that, in the traditional algorithm all the frequent attribute sets at each level starting from level 1 to level k are returned as the final sets of frequent attribute sets. This increases the number of attributes hugely, whereas in the proposed algorithm only the frequent attribute sets at level k are returned which are considered for materialization.

6.2 Proposed Algorithm

Input: $R(A_1, A_2, ..., A_m)$: The relation; $Q = \{Q_1, Q_2, ..., Q_n\}$: The set of user queries that will run on R; min_sup: minimum support, the static support count threshold, k: size of materialized view.

Output: M: The set of materialized views.

Step 1: Construct the m × n Attribute Usage Matrix (AUM) where m is the number of attributes and n is the number of queries.

Step 2: M = Apriori(AUM)

/* M is the set which stores the final sets of frequent k Attribute Sets in the kth level and Apriori is the method to invoke the Apriori algorithm and it takes AUM as the parameter.*/

Step 3: Execution of the Apriori Algorithm for generation of k-Attribute Set at the kth level only

Step 3.1: L_1 = find frequent 1-Attribute Set(AUM) /*frequent Attribute Set at level 1 is generated by taking AUM as the input */

Step 3.2: /* Step 3.2 is used to generate frequent attribute set at each level k starting from level 2*/

```
  If
   L₁=∅ then go to End.  /*Algorithm fails to generate any attribute(s) to be materia-
lized*/
   else
   for (k= 2; L_{k-1} !=∅; k++)
   do
   begin
   C_k= candidates generated from L_{k-1}; /* C_k: Candidate Attribute Set of size k, L_k: fre-
quent Attribute Set of size k */
       for each Query Q_i in AUM
         do
         increment the count of all candidates in C_k that are contained AUM
         L_k= candidates in C_k with count ≥ min_sup
       end for
   end for
```

Step 3.3: return L_k /* Returns the frequent Attribute Set(s) of size k*/

Step 4: Each member in M represents a materialized view having k number of attributes inside it.

Step 5: End.

7 Performance Evaluation of the Proposed Methodology Through Result Analysis and Comparison with Existing Methodologies

In this section a comparative study is presented between the proposed methodology and the techniques discussed in [6–8]. The key parameters in the evaluation of the Materialized views are (i) Query Hit-Miss ratio and (ii) The size of the Materialized View with respect to the size of the actual large database or fact table.

The implementation was carried out using .Net (visual Studio 2012) as front end and SQL Server 7.0 for the database and the operating system is Windows 7. For the experimental result we used an Intel core i3 3.1 GHz with the hard disk of 500 GB and RAM of 4 GB.

7.1 Result Analysis from Data Set-1

In order to analyze the performance of the methodologies proposed in [6, 8], the authors have executed the algorithms on a real life data. In this work the same dataset consisting of a student database of a state level examination system where the numbers of students is close to 1,00,000 and hence the numbers of tuples in the database.

The database schema of this examination system is given below.

T1: (Form_no, Name, Form_Type, Eng_Rank, Med_Rank, Category, PH)
T2: (Form_no, Name, F_name, DOB, Gender, Family_Income, Phn Number, School_Board)
T3: (Form_no, Name, Prefered_Coll1 (refers Col_id from T4), Prefered_Coll2 (refers Col_id from T4), Prefered_Coll3(refers Col_id from T4), Prefered_str1 (refers Streamid from T5), Prefered_str2 (refers Streamid from T5), Prefered_str3 (refers Streamid from T5))
T4: (Col_id, Col_Name, District, Year_of_Establishment)
T5: (Streamid, streamname)
T6: (Col_id(refers Col_id from T4, Streamid (refers Streamid from T5), Fees)

This analysis is done based on a recent query set. Based on the most frequently fired or executed queries we identify 10 mostly accessed attributes. The 10 most frequently accessed attributes are:

A_1 = Form_no, A_2 = Eng_Rank, A_3 = Name, A_4 = Form_Type, A_5 = Col_Name, A_6 = Col_Id, A_7 = Category, A_8 = District, A_9 = Streamid and A_{10} = . streamname

Due to space constraint only one of the important queries is mentioned below:

Q_3: select Col_Name, streamname from T_3, T_4, T_5 where T_3. Prefered_Coll1 = T_4. Col_Id and T_3. Prefered_str1 = T_5. Streamid and Form_no = 'VALUE';

Table 1 Data set-1

	A_1	A_2	A_3	A_4	A_5	A_6	A_7	A_8	A_9	A_{10}
Q_1	1	1	1	1	1	0	0	0	1	0
Q_2	1	0	0	1	0	1	1	0	0	0
Q_3	0	0	1	0	1	1	1	1	0	0
Q_4	1	0	0	0	1	1	0	0	1	1
Q_5	0	1	0	0	0	0	1	1	1	1
Q_6	0	0	1	1	0	1	0	0	0	1
Q_7	1	1	1	0	0	1	0	1	1	0
Q_8	1	1	1	0	1	0	0	0	0	1
Q_9	0	1	1	0	1	0	0	1	1	1
Q_{10}	1	0	0	1	0	1	1	0	1	1

From the above query it can be seen query Q_3 accesses the attributes A_1, A_5, A_6, A_9 and A_{10}, which is indicated in Table 1.

After identifying the most frequently accessed attributes by the most frequently executed queries a Query Attribute Relationship Matrix (QARM) (here denoted as Data Set-1) is formed in [6] to illustrate the working of their proposed algorithm based on liner regression method for construction of materialized view. The same QARM is used in [8] to illustrate the method to construct materialized view with Non Linear Regression. With the same QARM, the view created by the algorithm in [8] outperforms the view created by [6] in terms of Query Hit-Miss ratio.

The algorithms proposed in [6, 8] both give the flexibility to regulate the size of the materialized views according to the need of the organizations. One can increase or decrease the number of attributes present in a view by selecting the most important attributes. In our proposed technique at each level k, there would be k numbers of attributes inside a view. By varying the value of k, the number of attributes inside a view can be varied. Hence with the proposed technique also the size of the view can be monitored. In [6], two views are created from Data set-1. They are: V_1 (A_1, A_4, A_2) and V_2 (A_6, A_7). The view created in [8] from data set-1 is V_3 (A_6, A_9, A_3). V_1 and V_3 both contain three attributes, so to compare with both in this work we have created 3-Attribute set using Apriori algorithm.

The below table describes the highest level of Frequent Attribute Set that can be generated from Data Set-1 using Apriori algorithm with varying support count.

From Table 2 it can be seen when the support count is 30%, then highest level of attribute set that can be generated is three, i.e. 3-Attribute Set.

From Table 1 it can be seen total number of 3-Attribute Set generated with different combinations is four. The views are V_4 (A_1, A_2, A_3), V_5 (A_2, A_3, A_5), V_6 (A_2, A_3, A_9) and V_7 (A_2, A_8, A_9).

To compare the Hit-Miss ratio between different views, the following Query Frequency Matrix is considered (Table 3).

Table 2 Frequent attribute set from data set-1

Support count (%)	Highest level of frequent attribute set	Total number of combinations
20	4	7
30	3	4
40	2	4
50	1	7

Table 3 Query frequency matrix-1

Q_1	Q_2	Q_3	Q_4	Q_5	Q_6	Q_7	Q_8	Q_9	Q_{10}
21	8	27	4	23	8	8	23	22	6

From the above table it can be said that query Q1 is fired 21 times. As Q_1 accesses attributes A_1, A_2, A_3, A_4, A_5 and A_9. So for Q_1, the Hit values for the respective attributes are all 21.

7.2 Computation of the Hit-Miss Ratio

From Fig. 1 it can be seen that for the view V_1 created according to the methodology discussed in [6] the Hit-Miss ratio value is 37.3%. As claimed by the authors in [8] there is an improvement of almost 10% in Hit-Miss ratio for the view created by the methodology in [8]. Rest of the views, i.e. V_4, V_5, V_6 and V_7 are all created according to the proposed method. All of them outperforms the views created by

Fig. 1 Comparative study between existing methods and proposed work

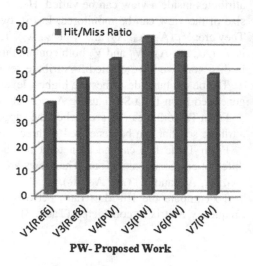

PW- Proposed Work

the methods discussed in [6, 8] in terms of Hit-Miss ratio. The highest Hit-Miss ratio value 64.5% is achieved for the view V_5 (A_2, A_3, A_5).

7.3   Result Analysis from Data Set-2

The work discussed in [7] constructs materialized views with a clustering based algorithm that measures the relationship among any two attributes with the help of a similarity function named Jaccard Index. The algorithm forms a set of clusters where every cluster denotes a materialized view. From result analysis it is seen that it performs very well in terms of query Hit-Miss ratio. The main drawback of this methodology is once a valid cluster is formed, the number of attributes inside the view is fixed. As the size of a materialized view is very critical, often the number of attributes inside a view needs to be changed. One can't increase or decrease the number of attributes inside a view with the methodology discussed in [7].

The authors in [7] have used an Attribute Usage Matrix to illustrate the working of their proposed methodology. In this work we have referred that matrix as Data Set-2. As discussed for Data Set-1, the AUM is constructed from a real life Sales Database. The database is designed according to relations mentioned in [23]. The schemas are given below:

T1: CLIENT_MASTER (clientno, name, address, city, pincode, state, baldue)

T2: PRODUCT_MASTER (productno, description, profitpercent, unitmeasure, qtyonhand, reorderlv, sellprice, costprice, product_master_pk)

T3: SALESMAN_MASTER (salesmanno, salesmanname, city, pincode, state, salamt, tgttoget, ttdsales, remarks)

T4: SALES_ORDER (orderno, clientno, orderdate, delyaddr, salesmanno, dely-type, billyn, delydate, orderstatus, constraint sales_order_con foreign key(clientno) references client_master (clientno), constraint so_frnkey_con foreign key (sales-manno) references salesman_master (salesmanno) enable)

T5: SALES_ORDER_DETAILS (orderno, productno, qtyordered, qtydisp, pro-ductrate, constraint sales_order_details_con foreign key (produtno) references product_master (productno), constraint sales_order_details_con2 foreign key (orderno) references sales_order (orderno))

As in Data Set-1, this analysis is done based on a recent query set. From the most frequently executed queries on this database we identify 12 mostly accessed attributes. The 12 most frequently accessed attributes are:

A_1 = clientno, A_2 = city, $A3$ = baldue, A_4 = name, A_5 = productno, A_6 = sellprice, A_7 = qtydisp, A_8 = orderno, A_9 = qtyordered, A_{10} = orderdate, A_{11} = productrate, A_{12} = orderstatus.

Among the 10 most frequently executed query on the sales database query Q_2 is mentioned below:

Table 4 Data set-2

	A1	A2	A3	A4	A5	A6	A7	A8	A9	A10	A11	A12
Q_1	0	0	1	1	1	0	1	0	1	1	1	0
Q_2	1	0	0	1	0	0	1	1	1	0	1	0
Q_3	0	1	1	1	1	1	0	1	0	1	0	1
Q_4	1	0	1	0	0	0	1	0	0	0	0	1
Q_5	0	1	0	0	1	1	1	1	1	1	0	1
Q_6	0	0	0	1	1	1	0	0	1	1	1	1
Q_7	1	1	1	1	0	1	0	0	1	0	1	0
Q_8	0	0	0	0	0	0	1	1	0	0	1	0
Q_9	0	1	0	0	1	0	0	1	0	1	0	0
Q_{10}	0	0	1	0	1	1	1	1	1	0	1	1

Q_2: select client_master.clientno as "clientno", client_master.name as "name", avg(sales_order_details.qtydisp) as "averagesales" from sales_order, sales_order_details, client_master where sales_order_details.orderno = sales_order.orderno and client_master.clientno = sales_order.clientno group by client_master.clientno, client_master.name having (sales_order_details.qtyordered * sales_order_details.productrate) > 15000.00

From the above query it can be seen query Q_2 accesses the attributes A_1, A_4, A_7, A_8, A_9 and A_{11}, which is indicated in the Attribute Usage Matrix (denoted as Data Set-2) (Table 4).

The views created in [7] are V_1 (A_5, A_{10}), V_2 (A_9, A_{11}), V_3 (A_6, A_{12}) and V_4 (A_4, A_5, A_6, A_9, A_{11}). We can see four views are created from the entire database which consists of twelve attributes. Three of them only have two attributes inside, which is only 16.66% and only one has five attributes inside which is almost 42% compared to the number of attributes present in the actual database.

The Table 5 describes the highest level of attribute set that can be generated from Data Set 2 with varying support count.

The decision makers of a company may want to place 50% of the entire database table attributes inside a view, so the number of attributes inside the view would be six. From Table 5 it can be seen when the support count is 20%, then highest level of attribute set that can be generated is six, i.e. 6-Attribute Set.

On execution of the proposed association based algorithm on Data set 2 with support count 20%, we get two views of size six. The views are V_5(A_2, A_5, A_6, A_8,

Table 5 Frequent attribute set from data set-2

Support count (%)	Highest level of attribute set	Total number of combination
20	6	2
30	4	4
40	3	2
50	2	2

Table 6 Query frequency matrix-2

Q_1	Q_2	Q_3	Q_4	Q_5	Q_6	Q_7	Q_8	Q_9	Q_{10}
25	63	27	42	38	51	67	44	49	72

Table 7 Hit-miss ratio based on no. of attributes in view

No. of attributes in a view	Hit-miss ratio
5	77
6	87

A_{10}, A_{12}) and $V_6(A_5, A_6, A_7, A_8, A_9, A_{12})$. As already mentioned the main drawback of the methodology discussed in [7] is it is unable to regulate the number of attributes in a view. $V_4(A_4, A_5, A_6, A_9, A_{11})$ is the highest sized view that can be generated by the methodology in [7] with size five.

To compare the Hit-Miss ratio between different views, the following Query Frequency Matrix is considered (Table 6).

Accordingly query Hit-Miss ratio for V_4 and V_5 have been calculated and depicted in Table 7.

From the above table the benefit of inserting one more attribute can seen here, where query Hit-Miss ratio increases considerably by 10%. This flexibility of resizing the view is not possible in the methodology discussed in [7].

8 Conclusion and Future Work

This research work proposes a methodology to construct materialized views from a large relation and a set of queries accessing the attributes of that relation. The existing works have constructed materialized views by considering the relationship or association between only a pair of attributes at a time. Hence to compute the association between n numbers of attributes the existing methods would split them into ($^nC_2$) number of pairs and calculate the association between each pair to transitively calculate the association between n number attributes. These approaches not only increase the computational complexities but also accuracy of the result has been compromised. In this research work a methodology using Association technique based Apriori algorithm has been proposed to form materialized views. The proposed algorithm can consider the association between more than two attributes together at a time. Thus the pair wise transitive attribute affinity measurement is avoided. From comparative result analysis it can be seen that the views formed by the proposed methodology outperforms the views formed by the existing methods in terms of both the key performance measuring parameters—query Hit/Miss ratio and view size scalability.

This research work can be further extended by considering the non binary data space instead of the binary data space on which the existing algorithms and the proposed Apriori based algorithm work. More accurate result is expected if non binary values instead of only '0' and '1' in the Attribute Usage Matrix (AUM) are considered. In real life scenario the queries are always generated from distributed environment having many nodes or sites. A further analysis can be carried out which will consider query frequency from different sites and accordingly the allocation of the most suitable views in proper sites to optimize the query processing in distributed environment.

References

1. Bazlur, A.N.M., Islam, R.M.S., Latiful Hoque, A.S.M.: Dynamic materialized view selection approach for improving query performance. Computer Networks and Information Technologies. Communications in Computer and Information Science, vol. 142, pp. 202–211 (2011)
2. Liu, Z., Chen, Y.: Answering keyword queries on XML using materialized views. In: IEEE 24th International Conference on Data Engineering (ICDE), pp. 1501–1503, Cancun, Mexico (2008)
3. Garnaud, E., Maabout, S., Mosbah, M.: Functional dependencies are helpful for partial materialization of data cubes. Springer J. Ann. Math. Artif. Intell. (2013)
4. Li, H., Huang, H., Liu S.: PMC: select materialized cells in data cubes. Data Warehousing and Knowledge Discovery. Lecture Notes in Computer Science, vol. 3589, pp. 168–178 (2005)
5. Sen, S., Dutta, A., Cortesi, A., Chaki, N.: A new scale for attribute dependency in large database systems. In: Springer LNCS Proceedings of the 11th International Conference on Information System and Industrial Management (CISIM), pp. 266–277, Venice, Italy (2012)
6. Ghosh, P., Sen, S., Chaki N.: Materialized view construction using linear regression on attributes. In: IEEE Proceedings of the 3rd International Conference on Emerging Applications of Information Technology (EAIT), pp. 214–219, Kolkata, India (2012)
7. Roy, S., Ghosh, R., Sen, S.: Materialized view construction based on clustering technique. In: 13th Springer-Verlag International Conference on Computer Information Systems and Industrial Management Applications (CISIM 2014), Ton Duc Thang University, Ho Chi Minh City, Vietnam, November 5–7, 2014
8. Sen, S., Ghosh, P., Cortesi, A.: Materialized view construction using linearizable non linear regression. In: 2nd International Doctoral Symposium on Applied Computation and Security Systems (ACSS), Department of Computer Science and Engineering, University of Calcutta, Kolkata, May 23–25, 2015
9. Aouiche, K., Jouve, P.: Clustering-based materialized view selection in data warehouses. In: Proceedings of 10th East European conference on Advances in Databases and Information Systems, pp. 81–95 (2006)
10. Harinarayan, V., Rajaraman, A., Ullman, J.: Implementing data cubes efficiently. In: Proceedings of ACM SIGMOD International Conference on Management of Data, pp. 205–216 (1996)
11. Bello, R.G., Dias, K., Feenan, J., Finnerty, J., Norcott, W.D., Sun, H., Witkowski, A., Ziauddin, M.: Materialized views in oracle. In: Proceedings of the 24th International Conference on Very Large Data Bases, pp. 659–664 (1998)
12. Vijay Kumar, T.V., Ghosal, A.: Greedy selection of materialized views. Int. J. Commun. Technol. 1(1), 156–172 (2009)
13. Chan, G.K.Y., Li, Q., Feng, L.: Design and selection of materialized views in a data warehousing environment: a case study. In: Proceedings of 2nd ACM International Workshop on Data Warehousing and OLAP, pp. 42–47 (1999)
14. Aouiche, K., Darmont, J.: Data mining-based materialized view and index selection in data warehouses. J. Intell. Inf. Syst. 33(1), 65–93 (2009)

15. Goswami, R., Bhattacharyya, D.K., Dutta, M., Kalita, J.K.: Approaches and issues in view selection for materialising in data warehouse. Int. J. Bus. Inf. Syst. **21**(1), 17–47 (2016)
16. Vijay Kumar, T.V., Kumar, S.: Materialized view selection using simulated annealing. In: Srinivasa, S., Bhatnagar, V. (eds.) BDA 2012, LNCS 7678, pp. 168–179. © Springer, Berlin Heidelberg (2012)
17. Vijay, T.V., Santosh Kumar, : Materialised view selection using randomised algorithms. Int. J. Bus. Inf. Syst. **19**(2), 224–240 (2015)
18. Das, A., Bhattacharyya, D.K.: Density-based view materialization. In: Pal, S.K., Bandy-opadhyay, S., Biswas, S. (eds.) Proceedings of First International Conference on Pattern Recognition and Machine Intelligence, PReMI 2005', LNCS, vol. 3776, pp. 589–594. Springer, Berlin (2005)
19. Kumar, T.V., Singh, A., Dubey, G.: Mining queries for constructing materialized views in a data warehouse. In: Wyld, D.C., Zizka, J., Nagamalai, D. (eds.) Advances in Computer Science, Engineering and Application, Proceedings of the Second International Conference on Computer Science, Engineering and Applications (ICCSEA 2012), vol. 2', volume 167 of Advances in Intelligent and Soft Computing, pp. 149–159. Springer, Berlin (2012)
20. Agarwal, R., Srikant, R.: Fast algorithms for mining association rules. In: Proceedings of the 20th International Conference on Very Large Data Bases, pp. 487–499 (1994)
21. Han, J., Pe, J., Yin, Y., Mao, R.: Mining frequent patterns without candidate generation: a frequent-pattern tree approach. Data Min. Knowl. Disc. **8**, 53–87 (2004)
22. Brin, S., Motwani, R., Ullman, J.D., Tsur, S.: Dynamic itemset counting and implication rules for market basket data. SIGMOD Record. **6**(2), 255–264 (1997)
23. Bayross, I.: SQL, PL/SQL the Programming Language of Oracle, 4th edn

A New Method for Key Author Analysis in Research Professionals' Collaboration Network

Anand Bihari and Sudhakar Tripathi

Abstract In research community, who are the most prominent or key authors in the research community is the major discussion or research issue. Different types of centrality measures and citation based indices are developed for finding key author in community. But main issues is what are the real contribution of an individual or group and their impact in research community. To find contribution of individual researcher, we use normalized citation count and geometric series to distribute the share to individual author in multi-authored paper. For evaluating the scientific impact of individual researcher, we use eigenvector centrality. In eigenvector centrality first, we set the initial amount of influence of each author to total normalized citation score and the collaboration weight is correlation coefficient value.

Keywords Social network · Research collaboration · Eigenvector centrality · Correlation coefficient

1 Introduction

In recent years, most of the research work is completed by two or more than two researchers. This collaboration prompts the productivity and production of new ideas. Researcher express these new ideas in the form of research articles and these ideas are studied by the other researchers and they will give the new ideas based on this. This is a process of carrying research work by the group of researchers. In group of researchers, individuals may be from the same or different subject areas and they form a network called research professionals collaboration network. In this network node represents the researcher and an edge between nodes represent the

A. Bihari (✉) · S. Tripathi
Department of Computer Science and Engineering,
National Institute of Technology Patna, Patna, Bihar, India
e-mail: anand.cse15@nitp.ac.in

S. Tripathi
e-mail: stripathi.cse@nitp.ac.in

© Springer Nature Singapore Pte Ltd. 2017   133
R. Chaki et al. (eds.), *Advanced Computing and Systems for Security*,
Advances in Intelligent Systems and Computing 567,
DOI 10.1007/978-981-10-3409-1_9

research collaboration. In this area, numerous research work is done for constructing a co-authorship network [1–3] and finding an influential person in the network based on graph theory, complex network, social network analysis and citation based index. Mainly co-authorship network is built according to their publication and the co-authorship weight, which is the total no of citation count of both researcher articles or total no of publication count published together. After that implement the social network analysis metrics as well as complex network properties to find key or prominent researcher in the network. In this paper, we discuss some of the related works which were done by the some of the eminent researcher and proposed a new method based on correlation of each pair of author. Finally we use eigenvector centrality to evaluate the scientific impact of an individual in research community.

2 Related Work

Newman [4] proposed a weighted system to construct a co-authorship network based on the number of co-author. The co-authorship weight between two author is the paper and their author ratio. After that use power law to calculate the total coauthorship weight between authors.

Farkas et al. [5] author discuss the weighted co-authorship network. In this network the collaboration weight are calculated based on the geometric mean.

Abbasia et al. [1, 6] author make a weighted co-authorship network and uses the social network analysis metrics for evaluating the performance of individual researcher. In this paper the collaboration weight is the total no. of paper published together.

Wang et al. [7] forms a smaller world of research professionals and investigated the co-author social network of HDM (Hierarchical Decision Model) research in SCI (Science Citation Index). In this paper Wang first construct the co-authorship network based on the publication details and consider the co-authorship weight is the total no. of publication and discusses the degree centrality, publication frequency and component analysis for finding key author in the network.

Liu et al. [8] investigated the co-authorship network of the Digital library (DL) Research community based on social network analysis metrics. Liu introduced a weighted directional network model to represent the co-authorship network of DL research community and proposed an author rank for individual authors and shows the advantage of author rank over social network analysis metrics (degree, closeness and betweenness centrality). In this paper author set the co-authorship is based on their paper and no. of coauthors. The co-authorship weight is the proportional of total no of author excluding own in a single article.

$$wc = \frac{1}{n-1} \tag{1}$$

Here wc indicates the co-authorship weight of author pair. The total collaboration weight of an author-pair cw (co-authorship weight) as follows:

$$cw = \sum_{i=1}^{n} wc_i \qquad (2)$$

Liu et al. [9] proposed a new method for constructing a co-author network. In this paper author evaluates the academic contribution based on their publication. Author suggested a method to give more weighted to the newly published papers rather than earlier published paper. Based on this principle author construct the paper citation network Gc, i.e. directed weighted network is denoted as Gc = (N, E, WC), where N is the set of papers, E is the set of edges represents the citation relationship between nodes and WC is total weight between node pairs I.e defines as follows:

$$WC_{ab} = \frac{\sigma_{ab}}{(Y_a - Y_b) - 1} \qquad (3)$$

where $\sigma_{ab} = 1$, if paper a cited by b otherwise 0. Y_a is the publication year of paper a and Y_b is the publication year of paper b. This function returns values between 0 and 1. If the paper is published earlier and cite later that means the paper has less importance in the network. So, the total score of an individual paper is sum of the weighted citation by all papers is defined as follows:

$$CC_a = \sum_{b=1}^{m} WC_{ab} \qquad (4)$$

The total share of an individual author in an article is calculated based on the geometric series. In this paper, author assumes that the first author gets the 1 share credit and rest of the author get k share credit based on geometric series. The share credit of ith author in a paper is calculated as follows:

$$S_{na} = \frac{(1-k) * k^{(R-1))}}{(1 - k^N A)} * CC_a \qquad (5)$$

where S_{na} depict the score of nth author. The R is the position of an author in a paper, NA is the total number of authors in the paper, k is the contribution of the first author above 30% = 0.7. The total score of an individual is the sum of all score of author obtained from each article. This method shows the influence of articles as well as individual authors. But in the co-authorship network needs to calculate the collaboration weight of two authors. For this author first calculate the effective distance between two authors based on the following equation:

$$D_{ab} = 1 - log(\frac{CT_{ab}}{CT_a}) \qquad (6)$$

where CT_{ab} is the collaboration times between author a and b and CT_a is the total publication of author a. After that calculate the co-authorship weight based on the law of gravity as follows:

$$W_{ab} = \frac{K * S_a * S_b}{d_{ab}^2} \tag{7}$$

where K is the constant, S_a, S_b are the total score of author i and j. W_{ab} is the collaboration weight between author a and b.

Bihari and Pandia [2] author evaluate the scientific impact of an individual scientist based on the social network analysis metric like degree, closeness, betweenness and eigenvector centrality and the citation based index like Frequency, Citation count, h-index, g-index and i10-index. For this, author downloads the publication details from IEEE and form a weighted undirected co-authorship network and the collaboration weight between two author is the total citation count of all papers that are published together.

Bihari and Pandia [10] author discuss the eigenvector centrality and its application in research professionals' relationship network and set the collaboration weight is the total citation count. For calculation of the eigenvector centrality author sets the initial amount of influence of each author is their degree centrality. After the calculate the eigenvector centrality value of each and every author. Finally, find out the prominent or key author in the network.

3   Methodology

3.1   Social Network Analysis (SNA)

SNA views social relationship between individuals in term of mathematics and graph theory, consisting nodes represents the individual actor, organizations or information in the network and edge represents the relationship like financial exchange, trade, friendship or sexual relationship exists between them [11–13]. In collaboration network a node represents a research professionals and link between two nodes represents that those authors are work together. SNA provides both visual and mathematical analysis of networks [14, 15].

3.2   Correlation Coefficient

It shows the cross correlation between two entities. The basic objective of correlation coefficient is to calculate the impact of correlation between two entities [16]. The correlation coefficient between any two entity is calculated as follows:

$$cr(a, b) = \frac{\sum_{k=1}^{m}(a_k - \overline{a})(b_k - \overline{b})}{(m - 1)St_a St_b}. \tag{8}$$

where a_k, b_k are the single unit, $\overline{a}$, $\overline{b}$ are the mean value of a and b, m is the total unit, St_a, St_b are the standard deviation of a and b.

3.3 Eigenvector Centrality

Eigenvector centrality is an important method to calculate the importance of a node with their neighbor's nodes and neighbor's nodes on their neighbors. It also gives the influence of a node. Basically it is based on degree centrality. In degree centrality simply count the direct connected neighbors but in this centrality it also consider the centrality value of neighbors or the importance of a node [17, 18]. If a node is connected with high eigenvector centrality nodes then they it may gain high centrality value than other with same number of neighbors with low eigenvector centrality because it carry the influence of neighbor's and their neighbor's nodes. The base of eigenvector centrality is the eigenvector and constant λ. Eigenvector centrality is proposed by the Philip Bonacichh [17] in 1987 and Google uses the concept of eigenvector centrality to rank the web pages is called PageRank. Let we consider a graph Gp = (N, L), where |N| represents the total number of nodes. Mathematically, eigenvector centrality is defined as follows:

$$EC_k = \frac{1}{\lambda} \sum_{p \in M(k)} EC_p = \frac{1}{\lambda} \sum_{p \in Gp} A_{k,p} EC_p \tag{9}$$

where M(k) is the set of the neighbors of node k, EC_p is the eigenvector centrality of node p, $A_{k,p}$ is the adjacency value of node pair of k and p and λ is a constant.

The essence of the eigenvector centrality is to the centrality score of a node is calculated based on their neighbors' node. If the neighbors' node having high centrality score and collaboration weight then have possibility to get high centrality value than others [19].

4   Data Collection and Cleansing

In order to evaluate the impact/influence of an individual scientist, first required to collect articles details. So, we collect the article details from IEEE Xplore [20] for the period of January 2000–July 2014 including journal and conferences proceedings. We can simply export the publication details from IEEE Xplore and it provides the publication data is in CSV(Comma Separated Value). The raw data contains several fields like Document title, Authors, Author's affiliation, Journal or conference name,

Table 1 Summary of data

Sl. no.	Attribute	Values
1	Data source	IEEE Xplore
2	Time period	January 2000–July 2014
3	No. of publication	26802
4	No. of journal	331
5	No. of conference	17487
6	No. of authors	61546
7	Total no. publication (Published in journal)	11220
8	Total no. of publication (Presented in conference)	15582

publication year, citation count etc. In order to construct a collaboration network, first extract the collaboration between two individual scientist based on the delimiter ";". After extraction of collaboration between two scientist, we found that the some of the scientist name was unable to read. So we replace this by the original name manually and replace the original name of an individual by unique number i.e. from 1 to 61546. After successfully cleaning of publication data we found 26802 articles and 61546 authors were available to conduct our experiments. First, we start the analysis of publication based on number of authors published a paper and another one is the collaboration strength of individual author. In our dataset we found that 2 paper are composed by the 100 or more than 100 authors i.e. 123 and 119. Three papers are composed by more than 50 authors, 36 papers are composed by 20 or more than 20 authors, 437 papers are composed by the 10 or more than 10 authors, 5167 papers are written by 5 or more authors, 4526, 6772, 7036, 3236 papers are composed by exact 4, 3, 2, 1 author(S) respectively. Author Lau, Y.Y. published 54 articles in different journal and conferences. Author Sasaki, M. published articles with 63 authors, around 17% author works individually and 38% author have collaboration with 2 person.The summary of data is shown in Table 1.

5 Collaboration Network of Research Professionals' an Example

For constructing a collaboration network, first we extract author and its co-author and their collaboration weight. The following techniques are used for extracting author, co-authors and collaboration weight. Let's, we use three papers: Art1, Art2 and Art3 details are shown in Table 2. After that we extract author and its co-author and the collaboration weight is the correlation coefficient [7, 21]. The network is constructed by using python and networkX [22]. The network is like Fig. 1.

Table 2 Example of publication and their authors

Paper ID	Author's name	Citation	Article type
Art1	A1, A2	12	Conference paper
Art2	B1, A3, B2	2	Conference paper
Art3	C1, A3, B2, C2, B1	11	Journal paper

Fig. 1 Research professionals' Collaboration network an example

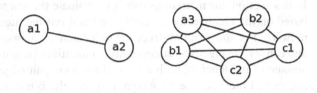

6 Our Proposed Method

Generally, authors used the collaboration weight is total number of publication [7] or total numbers of citation count of all papers that are published together [2] or total normalize citation citation count [8] or use law of gravity [9]. But, if an author-pair published 100 papers together but their citation count is less then we can only say that the author pair published more paper's than others but their collaboration impact is very less. If we consider the total number of citation count gained by together is a good measure of collaboration impact between these two authors. But suppose that the overall impact of one author is very good and the papers got citation based on the reputation of first author. So, it is not fair to consider the total citation count as a collaboration weight. Based on this limitation, we use correlation coefficient, to calculate the collaboration weight between the author-pair defined as follows:

$$CR(A1, A2) = \frac{(\overline{A1} - \mu_{(A1,A2)})(\overline{A2} - \mu_{(A1,A2)})}{N * S_{A1} * S_{A2}} \tag{10}$$

where $\overline{A1}$ and $\overline{A2}$ represents the mean citation value of author A1 and A2, $\mu_{(A1,A2)}$ represents the collaboration mean of author A1 and A2, N represents the total collaboration between A1 and A2 and S_{A1}, S_{A2} are the standard deviation of author A1 and A2 article's citation. In some cases the correlation coefficient gives the negative value, that means the collaboration between two authors has less impact than the individual one. The normalize citation count of each author is calculated based on Liu et al. [9]. In our approach, we consider the total share of first author is 50% = 0.5 instead of 0.7 used in [9]. Mathematically, the normalize citation count of an individual author is defined as follows:

$$NC_A = \sum_{i=1}^{n} NC_P \tag{11}$$

where NC_A is the total normalize citation count [9], n = Total publication counts and NC_P is the score of author in a particular paper [9].

7 Analysis and Result

In this section our main objective is to evaluate the scientific impact of individuals based on the eigenvector centrality and find out the most prominent author in the network. For this We construct the collaboration network based on available data and convey the eigenvector centrality. In traditional eigenvector centrality the initial amount of influence of each node and the node-pair edge weight is equal for every author and their pairs edge weight respectively. But in research community, every researcher have own importance and having difference with others and collaboration with co-author having also different with each others. Based on this limitation of traditional eigenvector centrality, we set the initial impact of each author to total normalized citation count [9] and the collaboration weight is the correlation coefficient.

We use python and networkX [22] as a simulation platform to compute the eigenvector centrality for all researchers in both measures proposed one and the traditional one. We select top 10 authors from both measures are shown in Tables 3 and 4 respectively proposed and traditional one.

Then we compare the result of top 10 authors from both measures (the comparative results is shown in Table 5) eigenvector centrality based on proposed model and tradition one and found that the proposed model gives better result than traditional eigenvector centrality and also no one author present in top 10 in both measures. In some cases the traditional eigenvector centrality gives better results than proposed one.

Table 3 Eigenvector centrality report of top 10 researcher in the network

No	Name of author	Eigenvector centrality
1	Romberg, J.	0.5777020637
2	Candes, E.J.	0.5772160296
3	Tao, T.	0.5769417419
4	Plan, Y.	0.0115131276
5	Frey, B.J.	0.0034618814
6	Loeliger, H.-A.	0.0034585634
7	Kschischang, F.R.	0.0034585634
8	Baraniuk, R.	0.0021154099
9	Hyeokho Choi	0.0021142049
10	Rodriguez, J.	0.0015333183

Table 4 Eigenvector centrality report of top 10 researcher based on traditional fashion in the network

No	Name of author	Eigenvector centrality
1	Mizuno, T.	0.1334739264
2	Kamae, T.	0.1333693085
3	Fukazawa, Y.	0.1295166699
4	Grove, J.E.	0.1288715982
5	Kuss, M.	0.1284210454
6	Schaefer, R.	0.1284210454
7	Ozaki, M.	0.1284210454
8	Dubois, R.	0.1284210454
9	Thompson, D.J.	0.1284210454
10	Lauben, D.	0.1284210454

Table 5 Comparative analysis of top 10 author

No	Name of author	Correlation coefficient based centrality value	Traditional centrality value
1	Romberg, J.	0.577702064	0.0396564
2	Candes, E.J.	0.57721603	0.080641074
3	Tao, T.	0.576941742	0.080641074
4	Plan, Y.	0.513127557	0.000642074
5	Frey, B.J.	0.346188139	0.000741074
6	Loeliger, H.-A.	0.24585634	0.039772202
7	Kschischang, F.R.	0.23585634	0.039685646
8	Baraniuk, R.	0.211540991	0.039675687
9	Hyeokho Choi	0.211420494	5.64003E-05
10	Rodriguez, J.	0.153331825	0.123331825
11	Schaefer, R.	0.15298185	0.128421045
12	Lauben, D.	0.15298185	0.128421045
13	Grove, J.E.	0.15298185	0.128871598
14	Mizuno, T.	0.149815294	0.133473926
15	Ozaki, M.	0.132981447	0.128421045
16	Kuss, M.	0.001531351	0.128421045
17	Kamae, T.	0.001529984	0.133369309
18	Dubois, R.	0.001529819	0.128421045
19	Thompson, D.J.	0.001529819	0.128421045
20	Fukazawa, Y.	0.001498153	0.12951667

8 Conclusion and Future Work

In this paper we have investigated the research professionals' collaboration network and find out the prominent actor in research community based on eigenvector centrality. For this first we set the initial amount of influence of each author to the normalized citation count and the collaboration weight is the correlation coefficient. We found that the proposed model gives better result than traditional one. In traditional eigenvector centrality, the initial amount of influence is the proportional of the total no. of the node and the collaboration weight is adjacency weight. But in proposed method the initial amount of influence of every node is normalize citation count that is based on the age of publication as well as the number of author. The normalize citation count is 1/(age of publication). But it penalize the old papers which got a significant amount of influence since their publication and also the distribution of citation with the authors is major research issue to give weightage to all authors. The limitation of proposed method is, it will require a more computation than traditional one.

References

1. Abbasi, A., Altmann, J.: On the correlation between research performance and social network analysis measures applied to research collaboration networks. In: 44th Hawaii International Conference on System Sciences (HICSS), 2011, pp. 1–10. IEEE (2011)
2. Bihari, A., Pandia, M.K.: Key author analysis in research professionals relationship network using citation indices and centrality. Proc. Comput. Sci. **57**, 606–613 (2015)
3. Pandia, M.K., Bihari, A.: Important author analysis in research professionals relationship network based on social network analysis metrics. In: Computational Intelligence in Data Mining, vol. 3, pp. 185–194. Springer (2015)
4. Newman, M.E.: Scientific collaboration networks. I. Network construction and fundamental results. Phys. Rev. E **64**(1), 016131 (2001)
5. Farkas, I., Bel, D., Palla, G., Vicsek, T.: Weighted network modules. New J. Phys. **9**(6), 180 (2007). http://stacks.iop.org/1367-2630/9/i=6/a=180
6. Abbasi, A., Hossain, L., Uddin, S., Rasmussen, K.J.: Evolutionary dynamics of scientific collaboration networks: multi-levels and cross-time analysis. Scientometrics **89**(2), 687–710 (2011)
7. Wang, B., Yao, X.: To form a smaller world in the research realm of hierarchical decision models. In: International Conference on Industrial Engineering and Engineering Management (IEEM), 2011 IEEE, pp. 1784–1788. IEEE (2011)
8. Liu, X., Bollen, J., Nelson, M.L., Van de Sompel, H.: Co-authorship networks in the digital library research community. Inf. Process. Manage. **41**(6), 1462–1480 (2005)
9. Liu, J., Li, Y., Ruan, Z., Fu, G., Chen, X., Sadiq, R., Deng, Y.: A new method to construct co-author networks. Phys. A Stat. Mech. Appl. **419**, 29–39 (2015)
10. Bihari, A., Pandia, M.K.: Eigenvector centrality and its application in research professionals' relationship network. In: 2015 International Conference on Futuristic Trends on Computational Analysis and Knowledge Management (ABLAZE), pp. 510–514. IEEE (2015)
11. Umadevi, V.: Automatic co-authorship network extraction and discovery of central authors. Int. J. Comput. Appl. **74**(4), 1–6 (2013)
12. Jin, J., Xu, K., Xiong, N., Liu, Y., Li, G.: Multi-index evaluation algorithm based on principal component analysis for node importance in complex networks. IET Netw. **1**(3), 108–115 (2012)

13. Wasserman, S., Faust, K.: Social Network Analysis: Methods and Applications, vol. 8. Cambridge University Press (1994)
14. Liu, B.: Web Data Mining. Springer (2007)
15. Said, Y.H., Wegman, E.J., Sharabati, W.K., Rigsby, J.T.: Retracted: social networks of author-coauthor relationships. Comput. Stat. Data Anal. **52**(4), 2177–2184 (2008)
16. https://www.mathsisfun.com/data/correlation.html
17. Bonacich, P., Lloyd, P.: Eigenvector-like measures of centrality for asymmetric relations. Soc. Netw. **23**(3), 191–201 (2001)
18. Newman, M.E.: The mathematics of networks. New Palgrave Encycl. Econ. **2**, 1–12 (2008)
19. Ding, D.-W., He, X.-Q.: Application of eigenvector centrality in metabolic networks. In: 2nd International Conference on Computer Engineering and Technology (ICCET), 2010, vol. 1, pp. V1–89. IEEE (2010)
20. http://ieeexplore.ieee.org/xpl/opac.jsp
21. Wang, B., Yang, J.: To form a smaller world in the research realm of hierarchical decision models. In: Proceedings of PICMET'11. PICMET (2011)
22. Schult, D.A., Swart, P.: Exploring network structure, dynamics, and function using networkX. In: Proceedings of the 7th Python in Science Conference (SCIPY 2008)

Part III
Security Systems

Single-Shot Person Re-identification by Spectral Matching of Symmetry-Driven Local Features

Aparajita Nanda and Pankaj K. Sa

Abstract This paper presents an appearance-based approach for person re-identification. It comprises the extraction of features that models the significant prospects of human appearance: local patterns of the regions and the presence of the spatial distribution of color, texture with high informative patches. This information is extracted from different body parts by following the symmetry-asymmetry precept and combined to form the signature of an individual. In this way, the signatures provide robustness against illumination variation, arbitrary pose changes, and occlusion. Further, these signatures are utilized for the process of re-identification through the spectral matching scheme. The approach applies to a situation where single image per individual is available. Extensive experiments are carried out on challenging datasets to validate the performance of proposed model over the state-of-the-art.

1 Introduction

Person re-identification deals with recognizing an individual who has previously been observed in diverse locations over disjoint camera views. It is considered as one of the crucial tasks in video surveillance, where consistent tracking and long-term activity have to be observed in a large structured area (airport, metro station). In this context, the solution like robust modeling of entire appearance of a person has been proved to be a better candidate for recognizing an individual, because the use of classical biometric traits (face, gait) may not be available due to poor camera resolution and crowded scenes. Usually, it is assumed that the attire of the individual

A. Nanda (✉) · P.K. Sa
Department of Computer Science and Engineering,
National Institute of Technology, Rourkela, India
e-mail: aparajita1.nanda@gmail.com

P.K. Sa
e-mail: pankajksa@nitrkl.ac.in

© Springer Nature Singapore Pte Ltd. 2017
R. Chaki et al. (eds.), *Advanced Computing and Systems for Security*,
Advances in Intelligent Systems and Computing 567,
DOI 10.1007/978-981-10-3409-1_10

remains same over the camera views. The model needs to be invariant against pose changes, viewpoint, and illumination [1]. These challenges demand specific appearance based solution.

In most of the approaches the person re-identification rely on the visual information and termed as appearance based approach [2–7]. However, other approaches modify the problem by adding temporal changes and spatial layout for matching the candidates [8, 9]. Schwartz et al. developed a model where the individuals are represented by a set of feature descriptors based on texture, color and shape [3]. Gheissari et al. addressed the spatiotemporal features with triangulation graph method for representing the spatial distribution of local features [9]. An unsupervised framework based on human salience or distinct features is developed and followed by feature matching [10]. The re-identification problem is formulated as distance metric learning problem [11–15]. The metric learning is based on the concept of maximizing the likelihood of true matches with relatively smaller distance than a false match pair. The re-identification is modeled as feature matching problem [2, 16–19]. The chromatic content, spatial arrangement and high structured patches are considered for matching the individuals [2]. Bak et al. proposed a model of covariance based human signature model that keeps the information of temporal changes [7]. Yang et al. discovers the salient color names based on salient color descriptor; it yields the assignment of higher probability value to the color names closer to the intrinsic color [20]. Zhao et al. design a learned discriminative mid-level filter from the cluster patches to recognize a specific visual pattern and distinguish the individuals [21]. Martinel et al. propose pairwise multiple metric learning frameworks to model different feature space individually [22]. The prerequisite of the training data is assumed as the main disfavor of learning based approaches. The feature matching requires minimal information content, by which global and local features are extracted from a single image and perform the matching against the gallery set candidates. In contrast, we systematically investigate the significance of extraction of salient features followed by a novel matching strategy for improving the accuracy.

In this paper, we present an appearance based person re-identification method by combining the local features. In the preprocessing step, the salient parts of the pedestrian image are selected from foreground by utilizing the symmetry-asymmetry principle that divides the body into head, torso, and leg. Crucial aspects of the local pattern of specific regions, the spatial distribution of color, texture are estimated through rgbSIFT and covariance descriptor respectively. The features are weighted with respect to the vertical axes computed from partitioning principle. The partitioning scheme deals with the minimization of arbitrary pose variations. While the rgbSIFT is considered to be robust against color variations and scale invariance. The spatial distribution of color, texture and illumination variations are handled by covariance descriptor. The spectral matching technique is employed for finding the matching between the features that yields the similarity measure between the individual.

The paper is organized as follows. Section 2 details the proposed approach. Experimental results are reported in Sect. 3 followed by conclusion in Sect. 4.

2 Proposed Approach

The proposed approach is a two-phase process. In the first phase, the features are extracted from the foreground regions of the pedestrian image and combined into a single signature. In the second phase, the correspondence signatures are determined by following the feature matching strategy. In the preprocessing step, the symmetry and asymmetry axes are computed for segmenting the pedestrian image.

2.1 Symmetry-Asymmetry Based Partitioning of Image

Symmetrical partitioning of an image aims to partition the pedestrian images into salient parts by applying the concept of symmetry and asymmetry principles. The detail description of the partitioning scheme is demonstrated in [2]. It divides each pedestrian image into two horizontal asymmetric axes on visual appearance and spatial information of the attire as described in Fig. 1. The images are resized into $W \times H$. The horizontal axes h are searched within the pedestrian image where $h_{H'T}$ separates the head and torso regions. The location of h_{TL} is searched within the region $[h - \delta, h + \delta]$, where it separates the regions with strongly distinct appearance between torso and leg (representing shirt/pants or suit/legs etc.). The value of δ is considered as $H/4$ that is relative to the height H of the pedestrian image. The head region is ignored because it contains few foreground pixels with less information content. So each pedestrian image is partitioned into two parts as R_k where $k = \{1, 2\}$ corresponds to torso and leg parts. Similarly, a symmetry y-axis w_{LR_k} for each R_1 and R_2 is computed that vertically divides the torso and leg regions. The body partition scheme divides the body part, according to the appearance of attire. This scheme is invariant to arbitrary changes of pose, different viewpoints and resolution.

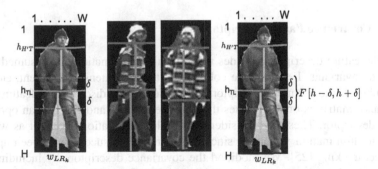

Fig. 1 Symmetrical partitioning of pedestrian image pairs. The two horizontal axis separate asymmetry regions such as head, torso and leg. The vertical axis are estimated for each symmetry region

Generally, the standard datasets for re-identification contain single-shot images as well as videos. In the case of single-shot, the Stel model [23] is customized for the separation of foreground and background. In the case of videos, background subtraction strategies are adequate to obtain the foreground part.

2.2 Accumulation of Local Features

Different local features are extracted from the detected torso and leg regions of the individual. The goal is to extract the heterogeneous information to compute the distinct signature model for the individuals. In order to minimize the pose variations, the features are considered with respect to the w_{LR_k} axes.

2.2.1 RgbSIFT

The SIFT descriptor depicts the local pattern of a particular region and operated on the gradient, which leads any changes to the gradient magnitude to remain ineffective. It means, despite the variations in light intensity of an image, still the gradient magnitude and directions remain invariant. However, the SIFT descriptor is not invariant to color changes. In order to provide invariant to color changes, the SIFT is applied to rgb channels individually. Thus rgbSIFT captures both the color and intensity transformations. The value of rgbSIFT descriptor is equivalent to the values of transformed color SIFT descriptor [24].

The SIFT descriptors are computed for each of symmetrical regions on the rgb channel independently. Thus, it results in the 128-dimensional feature vector for each key point over rgb channel respectively. The key points nearer to w_{LR_k} axes are the candidate key points which further take part in matching.

2.2.2 Covariance Patch Descriptor

Multiple feature descriptor provides more detailed information and assumed to be robust to invariants. Features like color, edge, texture, intensity, gradient, etc. can be combined into a single descriptor to provide a more discriminant feature. The covariance matrix is one that fuses the multiple features and can find an optimum global descriptor. These are considered as scale and rotation invariant as well as robust to illumination variations since this method had successfully been applied to object tracking [25]. We modified the covariance descriptors by including the features of color, edge, texture and shift invariance [26]. An image can be described by a set of covariance patch descriptors regardless of its dimensionality.

Let I be an image and Z be a $W \times H \times d$ dimensional feature vector extracted from I.

$$Z(x, y) = \Phi(I, x, y) \tag{1}$$

where the function ϕ can be any mapping function that is typified by a feature vector as follows

$$[x, y, H(x, y), S(x, y), V(x, y), G^i(x, y), \nabla^i(x, y), NFP(u, v)]^T \qquad (2)$$

where x, y denotes the pixel locations. $H(x, y)$, $S(x, y)$, $V(x, y)$ are the color features of HSV channels. $G^i(x, y)$, $\nabla^i(x, y)$ and $i \in [H, S, V]$ corresponds to gradient magnitude and orientation of each channel to extract the change in intensity and color of an image. Gradient magnitude and orientation are computed by convolving the image with a bank of log-Gabor filters. $NFP(u, v)$ represents the normalized Fourier coefficient that delineates the texture description of an image. NFP is obtained as

$$NFP(u, v) = \frac{|FP(u, v)|}{\sqrt{\sum_{(u \neq 0) \wedge (v \neq 0)} |FP(u, v)|^2}} \qquad (3)$$

where

$$FP(u, v) = T(P(x, y))$$

$FP(u, v)$ is the Fourier transform of pixel data $P(x, y)$. The advantage of using Fourier transform is that it posses shift invariance i.e. the transforms of a bit of large or uniform cloth segment will remain same. The magnitude transformed data is normalized by the sum of the squared values of each magnitude component. So that the normalized Fourier coefficient is invariant to linear shifts in illumination [27].

Each image is represented with sets of rectangular regions. The rectangular region $R' \subset Z$ and let z_b be the d-dimensional feature points within the region R'. Hence each of the regions R' is represented with $d \times d$ covariance matrix of the feature points

$$C = \frac{1}{n-1} \sum_{b=1}^{n} (z_b - \bar{z})(z_b - \bar{z})^T \qquad (4)$$

$\bar{z}$ is the mean feature vector. The covariance patch descriptor is a symmetric matrix, where the diagonal elements represent the variance of each feature, and the non-diagonal elements represent the correlation of features. To heighten the local illumination variation in an image, the covariance descriptors are normalized by using Pearson Product Moment Correlation Coefficients (PMCC). These normalized matrices are also termed as the correlation matrix. Afterward, every patch matrix is considered to be the normalized covariance patch matrix. The normalization is computed as

$$\overset{*}{C}(i, j) = \frac{C(i, j)}{\sigma_i \sigma_j} \qquad (5)$$

The $\frac{W}{4} \times \frac{W}{4}$ rectangular patches are extracted and shifted by $\frac{W}{8}$ horizontally and vertically within the image of the pedestrian image.

The spatial correlation between the patches signify the discriminating power of feature set. The covariance patches are the positive definite, and symmetric can be considered as tensors, and such defined tensor space is assumed as a manifold [28]. $\{C_i\}_{i=1}^{s}$ represents the set of normalized covariance patches extracted from each image.

Patch Homogeneity: Homogeneity of the covariance signifies the purity of the foreground pixels of the region with respect to color, gradient, and orientation. The idea is to filter out the most variable features because the features belonging to the background region are considered as the noisiest with less homogeneity value. The patch homogeneity is defined as

$$H(C_i) = \frac{1}{|C_i|} \sum_{z_i \in C_i} E(z_i, \bar{z}_i) \tag{6}$$

where

$$E(z_i, \bar{z}_i) = \left[1 - \frac{|z_i - \bar{z}_i|}{D\text{max}}\right]$$

$$D\text{max} = \max_{z_i \in C_i} \left(|z_i - \bar{z}_i|\right)$$

where $\bar{z}_i$ is the mean of the feature vector of the normalized covariance patch belongs to a specific region. $E(\cdot)$ is the distance measure of the feature vector from its mean feature vector. *Dmax* corresponds the maximum distance of any feature vector in the corresponding region to its mean. Thus, the patch with higher homogeneity values are selected for the Spectral matching (Sect. 2.3.1). In order to maintain the symmetries, the patches around the w_{LR_k}-axes are taken into consideration.

2.3 Feature Matching

This section describes the feature matching scores by combining the scores of rgb-SIFT and covariance feature descriptors. It illustrates the matching of signatures between the individuals belong to the gallery set G and probe set Q. The matching strategy may be of different types depending upon the content of the gallery and probe set.

The matching score of signatures of two individuals I_G and I_Q is estimated as:

$$\begin{aligned}
Score(I_G, I_Q) = \\
\beta_{rgbSIFT} \times d_{rgbSIFT}\left(rgbSIFT(I_G), rgbSIFT(I_Q)\right) \\
+ \beta_{cov} \times d_{cov}\left(Cov(I_G), Cov(I_Q)\right)
\end{aligned} \tag{7}$$

where β_i are the normalized weights and $i \in$ [rgbSIFT, Covariance patches]. $d_{rgbSIFT}$, d_{cov} are determined using spectral matching as explained in Sect. 2.3.1. In our experiment, we set the values of parameters as follows: $\beta_{rgbSIFT} = 0.4$ and $\beta_{cov} = 0.6$. These values are estimated from considering first 100 image pairs of VIPeR dataset and remain fixed for all experiments.

2.3.1 Spectral Matching

The purpose of spectral matching is to establish the pairwise relationship between the feature sets of probe and gallery respectively. It facilitates one to one as well as one of many matching strategies for the feature sets. Spectral correspondence is one of the robust techniques that require less computational complexity even for large datasets [29]. The spectral matching is modified for finding the matching score between the feature sets of covariance patches and rgbSIFT respectively.

Let us assume Q and G be the two feature sets of probe and gallery pedestrian image respectively. n_q and n_g are the number of key features in Q and G. The objective is to construct a weighted undirected graph A from the feature sets. A list L of n candidate assignment is formed, where the candidate assignments denote the corresponding mapping from one feature set to another one. For each candidate assignment $k = (i, i')$, $l = (j, j')$ where $i, j \in Q$ and $i', j' \in G$ affinity scores are computed. For pair of assignment (k, l) the affinity measures represents the compatibility between the features (i, j) and (i', j'). For every assignment $k \in L$ and every pair of assignment $(k, l) \in L$, a $n \times n$ affinity matrix A is formed where $A(k, k)$ represents the individual assignments from the candidate list L. The matrix A became $n \times n$ sparse symmetric matrix where $n = an_p$ and a is the average number of matching candidates for each feature $i \in Q$. All matching candidates are considered as positive link edge or cluster of assignments. A cluster Cl of assignments (i, i') is to be chosen that maximizes the inter cluster score C_s. A cluster is denoted with indicator vector I_v, where $I_v(k) = 1$ if $k \in Cl$ and else zero.

$$C_s = \sum_{k,l \in Cl} A(k, l) = I_v^T A I_v \tag{8}$$

In this case the problem becomes finding the cluster with the maximum inter-cluster score, that is resolved by eigenvector technique. The optimal solution I_v^* that maximizes the score as

$$I_v^* = \text{argmax}(C_s) \tag{9}$$

In the case of rgbSIFT, the spectral matching score between gallery and probe image is estimated independently for each of the corresponding regions. The overall score between a gallery and probe image is computed by averaging the scores of all the corresponding partitions and normalized by the maximum matching scores among all the symmetric regions. However in the case of covariance, the patches with high homogeneity near w_{LR_k} axis are considered as the feature sets for matching.

The computed patch homogeneity value defines the affinity score for the individual assignment. The covariance patches of a particular region with size $b \times d \times d$ are modeled as $b \times d^2$, where b is the number of covariance patches and $d \times d$ denotes the covariance patch of a region.

3   Experimental Evaluation

Section 3.1 presents the qualitative results and comparisons of single-shot evaluation on the datasets. We investigate the re-identification execution and the effects of the combination of features at different levels in Sect. 3.2.

Datasets: There exist challenging datasets ETHZ [3], VIPeR [6] which are used for person re-identification. These datasets provide various challenging prospects for person re-identification problem such as arbitrary pose changes, different viewpoints, occlusions and illumination variations, etc.

- The images of the ETHZ dataset are captured by moving the camera and offers variations in appearances of the image. All the images of this dataset are normalized to 64×32 pixels. It is structured as three sequences of pedestrian images such as ETHZ Seq#1 contains 83 pedestrian images, ETHZ Seq#2 contains 35 pedestrian images, and ETHZ Seq#3 contains 28 pedestrian images. Illumination variations and occlusion are the challenging issues of ETHZ dataset.
- The VIPeR dataset consists of 632 pedestrian image pairs taken from two camera views. It contains the pair of images for each individual from disjoint cameras. The images of VIPeR datasets are normalized up to 48×128 pixels.

Evaluation Criteria: The recognition rates are evaluated with the Cumulative Matching Characteristic (CMC) curves [6]. The CMC curve represents the true matches in top ranks with respect to the recognition rates. The recognition rates are computed from the average results of 10 different runs of test sets for both the datasets.

3.1   Single-Shot (SvsS)

Concerning the single-shot case of VIPeR dataset, the images captured on one camera are defined as gallery set and the images from another camera are considered as the probe set. The dataset is divided into the training set and testing set followed by the matching scheme. The dataset is partitioned into 316 and 474 pedestrian testing sets. The proposed approach is compared with the existing techniques such as symmetry-driven accumulation of local features (SDALF [2]) with 10 runs, Rank SVM (PRSVM [30]) with 5 runs, Viewpoint invariant pedestrian recognition (ELF [6]) with 10 runs, and the average scores are considered. In order to compare fairly, the proposed approach with other existing approaches necessitates the splitting

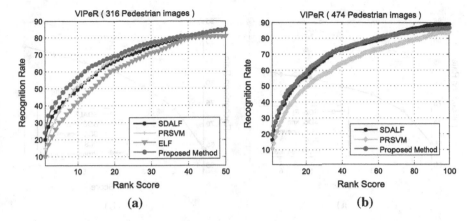

Fig. 2 Recognition rates on the VIPeR dataset in terms of CMC scores. In **a** and **b** the performance of proposed method is compared with SDALF [2], ELF [6] and PRSVM [30] on VIPeR dataset with 316-pedestrian and 474-pedestrian images, respectively

of datasets. However the information is not noted in the former works, so we compare the results of existing methods with the average results of proposed approach. The average results are computed from 10 different test sets of 50% (316) and 75% (474) pedestrian images of VIPeR datasets.

Figure 2 presents the results of proposed approach in terms of CMC. The performance of proposed approach is compared with the existing single-shot approaches on VIPeR dataset such as SDALF, PRSVM, ELF. From Fig. 2a it can be observed that the proposed approach outperforms all the other approaches, whereas the recognition rates of PRSVM and SDALF are worthy enough and nearly same with a difference of 0.12%. Figure 2b describes the result of proposed approach for 474 test images and compared with PRSVM and SDALF. The proposed method outperforms the PRSVM for the first ranks in CMC curve. This indicates that in a real scenario the performance of proposed approach provides better recognition results.

The results of the single-shot case for ETHZ datasets are reported in Fig. 3. The performance of proposed approach is compared with the existing single-shot approaches on ETHZ, such as PLS [3], SDALF [2] and salience learning based approach SDC_knn, SDC_ocsvm [10].

3.2 Accumulation of Features

This section describes the individual performance of the local features such as covariance patches, rgbSIFT as well as effectiveness when these are combined. The recognition are demonstrated on both VIPeR and ETHZ datasets. VIPeR dataset with 316 test images (Fig. 4a), the top rank average recognition results covariance

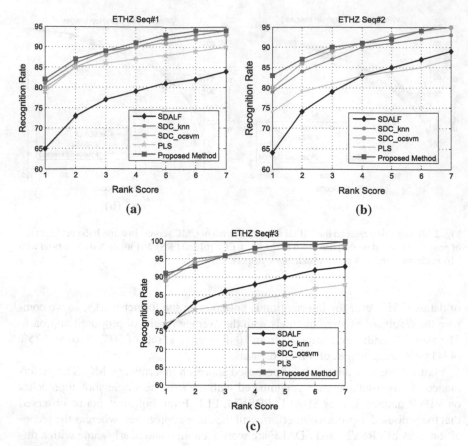

Fig. 3 The **a**, **b** and **c** represents the recognition rate for ETHZ Seq#1, Seq#2 and Seq#3 datasets respectively.

and rgbSIFT are 8 and 10% respectively. However, the combination rgbSIFT and covariance increases the recognition rates to 20%. Instead for VIPeR with 474 test images (Fig. 4b) the combination of rgbSIFT and covariance patch yields 19.21% top ranking recognition rates. The images of VIPeR datasets deals with viewpoint and illumination variations that can be counterbalanced by the integration of two features.

Table 1 presents the essence of rgbSIFT and covariance patches and also determine the recognition rates with respect to the combination of features. The comparison of performance results with respect to the combination of features on ETHZ datasets are also demonstrated. From Table 1, it can be observed that the addition of more information to the signature yields better recognition rates.

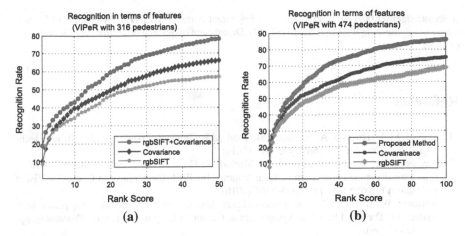

Fig. 4 The performance of rgbSIFT and covariance features and the fusion of these features on the VIPeR dataset in terms of CMC. In **a** and **b** the recognition rates are compared with the performance of individual feature descriptors as well as the combinations of the features on VIPeR dataset with 316-pedestrian and 474-pedestrian images, respectively

Table 1 The quantitative comparison of the recognition rates in percentage for the accumulation of feature descriptors such as rgbSIFT, and Covariance patches on ETHZ Seq#1, ETHZ Seq#2, ETHZ Seq#3 datasets

	ETHZ Seq#1				ETHZ Seq#2				ETHZ Seq#3			
Ranks(r)	1	3	5	7	1	3	5	7	1	3	5	7
rgbSIFT	76.27	82.43	86.40	88.81	79.41	83.25	85.46	87.34	89.32	92.35	94.38	96.13
Covariance	79.34	84.52	87.51	89.54	80.31	84.55	86.47	88.32	90.27	94.28	98.14	98.26
rgbSIFT+ Covariance	**81.91**	**89.24**	**92.87**	**94.56**	**83.03**	**89.92**	**92.62**	**96.15**	**92.03**	**96.33**	**98.52**	**99.51**

4 Conclusions

In this paper, an appearance-based approach for person re-identification has been presented that includes feature extraction followed by matching scheme. The approach adopts the symmetry and asymmetry precept for the partitioning of the pedestrian image that handles the viewpoint invariance. Crucial information such as the local pattern of regions and spatial distribution of color and texture descriptor are extracted. This provides the hardiness to arbitrary pose changes, illumination variations, and scaling. The use of spectral matching scheme gains, improvement in the matching rates. It is seen that its performance in terms of recognition rate is commendable when compared with its counterparts. This approach works for single-shot. However, it can be extended for multi-shot modality by adding more flexibility.

Acknowledgements This work is supported by Grant Number SB/FTP/ETA-0059/2014 by Science and Engineering Research Board (SERB), Department of Science & Technology, Government of India.

References

1. Saghafi, M.A., Hussain, A., Zaman, H.B., Saad, M.H.: Review of person re-identification techniques. IET Comput. Vis. **8**(6), 455–474 (2014)
2. Farenzena, M., Bazzani, L., Perina, A., Murino, V., Cristani, M.: Person re-identification by symmetry-driven accumulation of local features. In: IEEE Conference on Computer Vision and Pattern Recognition, pp. 2360–2367 (2010)
3. Schwartz, W., Davis, L.: Learning discriminative appearance-based models using partial least squares. In: The 22nd Brazilian Symposium on Computer Graphics and Image Processing, pp. 322–329 (2009)
4. Hirzer, M., Roth, P., Ostinger, M.K., Bischof, H.: Relaxed pairwise learned metric for person reidentification. In: Computer Vision-ECCV, pp. 780–793 (2012)
5. Zheng, W.S., Gong, S., Xiang, T.: Associating groups of people. In: British Machine Vision Conference (2009)
6. Gray, D., Tao, H.: Viewpoint invariant pedestrian recognition with an ensemble of localized features. In: Computer Vision–ECCV, pp. 262–275 (2008)
7. Bak, S., Corvee, E., Bremond, F., Thonnat, M.: Boosted human re-identification using riemannian manifolds. Image Vis. Comput. **30**(6), 443–452 (2012)
8. Javed, O., Shafique, K., Rasheed, Z., Shah, M.: Modeling inter-camera space-time and appearance relationships for tracking accross non-overlapping views. CVIU Elsevier **109**(2), 146–162 (2008)
9. Gheissari, N., Sebastian, T.B., Hartley, R.: Person reidentification using spatiotemporal appearance. In: IEEE Conference on Computer Vision and Pattern Recognition, vol. 2, pp. 1528–1535 (2006)
10. Zhao, R., Ouyang, W., Wang, X.: Unsupervised salience learning for person re-identification. In: IEEE Conference on Computer Vision and Pattern Recognition (CVPR), pp. 3586–3593 (2013)
11. Zheng, W.S., Gong, S., Xiang, T.: Reidentification by relative distance comparison. IEEE Trans. Pattern Anal. Mach. Intell. **35**(3), 653–668 (2013)
12. Mignon, A., Jurie, F.: Pcca: A new approach for distance learning from sparse pairwise constraints. In: 2012 IEEE Conference on Computer Vision and Pattern Recognition (CVPR), pp. 2666–2672 (2012)
13. Li, W., Zhao, R., Wang, X.: Human reidentification with transferred metric learning. In: Computer Vision-ACCV, vol. 2012, pp. 31–44 (2013)
14. Dikmen, M., Akbas, E., Huang, T.S., Ahuja, N.: Pedestrian recognition with a learned metric. In: Computer Vision-ACCV, vol. 2010, pp. 501–512 (2011)
15. Wu, Y., Minoh, M., Mukunoki, M., Lao, S.: Set based discriminative ranking for recognition. In: Computer Vision-ECCV, vol. 2012, pp. 497–510 (2012)
16. Satta, R., Fumera, G., Roli, F., Cristani, M., Murino, V.: A multiple component matching framework for person re-identification. In: Image Analysis and Processing-ICIAP, pp. 140–149 (2011)
17. Cheng, D.S., Cristani, M., Stoppa, M., Bazzani, L., Murino, V.: Custom pictorial structures for re-identification. In: BMVC, vol. 2, issue no. 5 (2011)
18. Wang, X., Zhao, R.: Person Re-identification: System Design and Evaluation Overview, vol. 2, pp. 351–370. Springer (2014)
19. Hamdoun, O., Moutarde, F., Stanciulescu, B., Steux, B.: Person re-identification in multi-camera system by signature based on interest point descriptors collected on short video

sequences. In: Second ACM/IEEE International Conference on Distributed Smart Cameras, 2008. ICDSC 2008, pp. 1–6 (2008)

20. Yang, Y., Yang, J., Yan, J., Liao, S., Yi, D., Li, S.Z.: Salient color names for person re-identification. In: Computer Vision–ECCV 2014, pp. 536–551. Springer (2014)

21. Zhao, R., Ouyang, W., Wang, X.: Learning mid-level filters for person re-identification. In: IEEE Conference on Computer Vision and Pattern Recognition (CVPR), pp. 144–151. IEEE (2014)

22. Martinel, Niki, Micheloni, Christian, Foresti, Gian Luca: Kernelized saliency-based person re-identification through multiple metric learning. IEEE Trans. Image Process. 24(12), 5645–5658 (2015)

23. Jojic, N., Perina, A., Cristani, M., Murino, V., Frey, B.: Stel component analysis: Modeling spatial correlations in image class structure. In: IEEE Conference on Computer Vision and Pattern Recognition, 2009. CVPR 2009, vol. 30, issue no. 6, pp. 2044–2051

24. Sande, K.V.D., Gevers, T., Snoek, C.G.: Evaluating color descriptors for object and scene recognition. IEEE Trans. Pattern Anal. Mach. Intell. 32(9), 1582–1596 (2010)

25. Tuzel, O., Porikli, F., Meer, P.: Region covariance: a fast descriptor for detection and classification. In: Proceedings of the 9th European Conference on Computer Vision (ECCV), pp. 586–3593 (2006)

26. Nanda, A., Sa, P.K., Majhi, B.: Covarinace based person re-identification using spectral matching. In: INDICON (2014)

27. Nixon, M., Aguado, A.: Feature Extraction & Image Processing. Academic Press (2008)

28. Pennec, X., Fillard, P., Ayache, N.: A riemannian framework for tensor computing. Int. J. Comput. Vis., 41–66 (2006)

29. Leordeanu, M., Hebert, M.: A spectral technique for correspondence problems using pairwise constraints. In: Proceedings of 10th IEEE International Conference on Computer Vision (ICCV), pp. 1482–1489 (2005)

30. Prosser, B., Zheng, W.S., Gong, S., Xiang, T.: Person re-identification by support vector ranking. In: British Machine Vision Conference, vol. 2, issue no. 5, pp. 21.1–21.11 (2010)

Analysis of Eavesdropping in QKD with Qutrit Photon States

Supriyo Banerjee, Biswajit Maiti and Banani Saha

Abstract Quantum Key Distribution (QKD) is used for generating key between two legitimate user Alice and Bob. In the past decade, QKD using qubit has been investigated for different cryptographic strategies. Analysis of qutrit photon states in QKD is theoretically investigated and different eavesdropping strategies have been discussed in this paper. Modified two stage Positive Operator Value Measurement (POVM) is used to determine the qutrit states. An expression for error rate in data transmission has been formulated both in case of entangled and opaque eavesdropping. A protocol for encryption of data in qutrit states in one-time-pad scheme is initiated to improve on information reconciliation.

Keywords Quantum cryptography · Qutrit · POVM · One-time-pad scheme · Entangled translucent eavesdropping

1 Introduction

In the beginning of last decade when it has been realized that nanotechnology has been matured to such an extent that single photon state can be generated and detected in the new kind of quantum dot (QD) lasing devices and detectors, people have tried to develop more secured quantum cryptographic techniques utilizing

S. Banerjee (✉)
Department of Computer Science, Kalyani Government Engineering College,
Kalyani 741235, India
e-mail: sban59@yahoo.com

B. Maiti
Department of Physics, Darjeeling Government College, Darjeeling 734101, India
e-mail: bmkgec@gmail.com

B. Saha
Department of Computer Science and Engineering, University of Calcutta,
Kolkata 700009, India
e-mail: bsaha_29@yahoo.co.in

© Springer Nature Singapore Pte Ltd. 2017   161
R. Chaki et al. (eds.), *Advanced Computing and Systems for Security*,
Advances in Intelligent Systems and Computing 567,
DOI 10.1007/978-981-10-3409-1_11

polarized photons where information is encoded in photons having specific phase relations as fixed by the users. Several schemes have been developed for key encryption with single and multi photon polarization states with orthogonal and non-orthogonal bases and their experimental realization has been partially verified. Also the security aspects of these schemes are under intensive investigation. Ekert et al. [1] were among the pioneers to present a thorough discussion on one such quantum communication protocol between two legitimate users in presence of possible eavesdropping strategies. They have developed the protocol with two equally probable non-orthogonal polarization states of a single photon $\langle u|$ and $\langle v|$ encoded as 0 and 1 for the purpose of secure key generation. In the above mentioned paper they have thoroughly analyzed errors in communication in three possible eavesdropping cases namely (i) Opaque (ii) Translucent and (iii) Entangled Translucent.

In the scheme of eavesdropping, interference or other measurement techniques may be employed by Eve in the beam path to extract information, thereby the data that will reach the receiver will be a modified one depending on the outcome of the measurement result of Eve. This is the case of opaque eavesdropping and the data transmission is completely erroneous. The presence of Eve in this case is easily detected by Bob from his results of detection. Unless correct basis is known quantum uncertainty forbids correct state determination by Eve [2] and hence Bob's data become erroneous because of intervention. Therefore, such kind of data transmission has to be discarded. Since quantum measurement is sure to expose her presence Eve on her stride to maximize information may use POVM technique as another strategy. In POVM, she allows the polarized photon to interact unitarily with her probe photon and then lets it to proceed to Bob thereby slightly modifying the original polarization state of the photon as a scheme of translucent eavesdropping. In doing so, she actually increases the overlap between the two photon states sent to Bob thereby reducing the exchange of information between Alice and Bob [1]. This scheme hides her presence but information extraction by Eve is comparatively poor. So in order to increase the mutual information sharing with Bob, she should try to entangle her probe state with the changed states of photons which will be forwarded to Bob. In this case of entangled translucent eavesdropping since Bob receives a mixed state of photons so realization of key and extraction of information will be difficult. Eve may attach her probe to all the bits or a fraction of bits up to her satisfaction. The first scenario is termed as collective attack. Brandt et al. [3] in their article have discussed some aspects of entangled translucent eavesdropping with Bennett's two state protocol with two non-orthogonal states of photon [4], where they have considered that Eve makes information maximizing Von Neumann type of projective measurement and Bob performs POVM. Since, non-orthogonal operators are non-commutative, so mathematical non-commutative operators representing measurement of non-orthogonal polarization states may provide a way of theoretical analysis for the detection of any kind of eavesdropping. Also because it is difficult to clone [5] arbitrary quantum states, so in principle, eavesdropping can easily be detected in Bennett's two state protocol. On the other hand, an extensive quantum theoretical analysis of eavesdropping strategies with

POVM is presented by Fuchs and Peres [6, 7]. All this analysis shows that the more disturbance Eve can create, the more information she will share with Bob. But to shield the information sharing between Bob and Eve we have tried to embed the encoded key in the main data string encapsulated by fencing states. For this we make use of the third polarization state of a qutrit to give Bob the information about the location of key embedded in the main data string, while hiding it to Eve. Also, the fence position or the length of the string containing the information about key is being changed in one time pad scheme in different successive communications. It is very difficult for Eve to keep track of it.

In this protocol, two of the non-orthogonal states of a qutrit have been used for encoding key where impurity count in the cipher text is the key while the third state provides the demarcation between the data and the location of the embedded key. This modification of Bennett's two state protocol have been presented in Sect. 2.

The modified design of POVM to identify the qutrit states has been discussed in Sect. 3. Probabilistic analysis has also been made in this section to obtain the closed form expression for the rate of mutual information exchange and the error rate in quantum communication channels as a function of POVM receiver error rate and the angle between the carrier polarization states.

Section 4 concludes the paper with the analysis of the results described in Sect. 3.

2 Qutrit Protocol

In this protocol, Alice prepares a string of random bits in either of the three non-orthogonal polarization states $|u\rangle$, $|v\rangle$ and $|w\rangle$ of a qutrit. Two of the states $|u\rangle$ and $|v\rangle$ are used to prepare embedded key and the cipher data will be the mixed quantum states of $|u\rangle$ and $|v\rangle$, where $|v\rangle$ is taken as the impurity in $|u\rangle$; the number of impurity count will be the key. Then the encoded key is sent to Bob through a quantum channel. One time pad procedure i.e., changing the data length time to time, depending on the number of impurity, is used to provide additional secrecy. Since key is embedded in a small section of the main data string, it is not possible to extract key merely from its number count, rather one must know the exact location where it is embedded. We have used the third state $|w\rangle$ to represent the line of fencing between the body of data and impurity counter. So the cipher data length will be N + 2L + 1 where N is the number of bits carrying actual information and L is the number of impurities. As we have employed one time pad scheme, in every successive communication the length of L will keep on changing providing much better security towards its determination by any third party, Eve.

After receiving a data, which is a photon in a particular polarization state, Bob in his POVM first detects it in randomly selected basis and discusses the result with Alice via public channel to determine the fence trit. Following the response from Alice, Bob chooses another basis, again discusses the detection result with Alice. Depending upon the agreement or disagreement, he follows the process 2n times up to his satisfaction and then fixes the computational basis to determine the fence trit. After determining the fence position, Bob with the selected computational basis detects and decodes the key by counting the number of impurities from the ciphered text. He starts counting impurity from the right most position, forms the key and then decrypts the data using that Key. If there is no agreement after 2n such discussions, they do decide to discard data. Since key is determined from the count of number of impurities in the ciphered text before the fence and equal number of impurity photons after the fence, this will provide a self checking mechanism at Bob's end. So once the fence position is settled, Bob at his end can detect data error, if any, through comparison and thereby can extract key.

3 POVM Receiver

The POVM receiver as designed by Brandt et al. [3] for the determination of two non-orthogonal states of a qubit has been modified to determine the three non-orthogonal polarization states of a qutrit. It is basically a two stage POVM receiver. In the first stage, Bob detects $|w\rangle$ by tracing out the combined state $|\phi\rangle$ of $|u\rangle$ and $|v\rangle$ in consultation with Alice of his detection results. As $|w\rangle$ is detected in the first stage, it is sure that the state $|\phi\rangle$ will not appear at the polarization dependent beam splitter P2 of the second stage of POVM, otherwise $|\phi\rangle$ has a finite probability to be at P2. In the second stage, same technique is employed to detect the states $|u\rangle$ and $|v\rangle$ by tracing out one or the other. Since $|w\rangle$ is the state that encapsulates the positions of key in the cipher text, the count of $|v\rangle$ in $|u\rangle$ between two successive $|w\rangle$ states will provide the key. The data string is designed in such a way that the count of $|v\rangle$ in $|u\rangle$ before the detection of fence state $|w\rangle$ will be equal to the number of $|v\rangle$ states till next $|w\rangle$ is detected. It therefore provides a self verifying mechanism that establishes the key.

For realization of the protocol, the vector space of the photons is projected into the orthogonal vector space spanned by the qutrit with the help of two polarization dependent beam splitters P_1 and P_2 at the two stages of POVM. The resulting states will be the mixed states of the original photon states depending upon their phase relations with the orthogonal state vectors. Afterwards, they are interferometrically recombined to get back the original states subject to the realization of the ambiguous result. Then at the end of communication and subsequent detection, Alice and Bob through public discussion estimate the rate of inconclusive results R. If it is beyond some expected value R_0, Bob will look for eavesdropping. Otherwise, within the limit of discrepancy, Bob estimates the error rate $Q = \frac{q}{(1-R)}$, where q is the probability of data transmission error before discarding the inconclusive results. Non-zero value for Q, in general, indicates eavesdropping. So for a given error rate Q, their task is to estimate how much information is leaked to Eve and to decide if the data on hand can be corrected and be used to extract key [8].

As the photon is in any one of its three arbitrary polarization states in the computational basis, it can be written as the linear combination of these three states

$$|\psi\rangle = e^{i\chi_1}a_0|u\rangle + e^{i\chi_2}b_0|v\rangle + c_0|w\rangle \tag{1}$$

where a_0, b_0, c_0 are the real constants representing the relative contribution or relative count of the respective polarization states in the data string with respect to the computational basis. Upon suitable choice of phases and basis, the overlaps between any two photon states can be represented by the angles θ and φ: $\langle u|v\rangle = \cos\theta$, $\langle v|w\rangle = \sin\theta\sin\varphi$ and $\langle u|w\rangle = \cos\theta\sin\varphi$ apart from the phase factors χ_1 and χ_2 of $|u\rangle$ and $|v\rangle$ with respect to $|w\rangle$. These phase factors are introduced to encode the photons for the purpose of key generation.

Whenever Alice and Bob communicate through binary erasure channel, Bob may sometimes get ambiguous results even n the absence of any eavesdropper. Once Bob recognizes a particular photon in its correct state of polarization, the other one will be the definite one to him. This is the advantage of POVM in reducing the probability of inconclusive results [9]. The positive value for any detector A_w indicates that the detected photon is in state $|w\rangle$ and so are the cases for other detectors. A null result may give ambiguous conclusion.

Since in the first stage Bob has to detect $|w\rangle$, the corresponding measurement operator can be set as

$$A_w = \frac{P_w}{1+S} \quad \text{where } P_w = (1 - |\phi\rangle\langle\phi|), \quad S = \langle\phi|w\rangle.$$

$|\phi\rangle = a_0|u\rangle + e^{i(\chi_1 - \chi_2)}b_0|v\rangle$ is the resultant of $|u\rangle$ and $|v\rangle$ polarization states. If the angle between $|w\rangle$ and $|\phi\rangle$ be φ then $\langle\phi|w\rangle = \cos\varphi$. Then the detection of $|w\rangle$ results in $\langle w|A_w|w\rangle = (1 - \cos\varphi)$ .Hence the probability of getting ambiguous result is $\langle w|A_?|w\rangle = \cos\varphi$. But actually the photon is in the general quantum state $|\psi\rangle$ representing any of the three qutrit states so the detection of $|w\rangle$ will result in some lower probability corresponding to its extent of presence in the data string and will be $\langle\psi|A_w|\psi\rangle = c_0^2(1 - \cos\varphi)$ while the detection probability of state $|\phi\rangle$ will be $\langle\psi|A_\phi|\psi\rangle = (a_0 + \zeta b_0)^2(1 - \cos\varphi)$ where ζ accounts for the relative phase contribution between $|u\rangle$, $|v\rangle$ states and introduces a sign factor or determines the polarization state of $|\phi\rangle$. Since the ambiguous result is $\langle\psi|A_?|\psi\rangle = (a_0 + b_0 + c_0)^2\cos\varphi$ then to satisfy the unitarity criterion, following relation must hold from which phase relation between $|u\rangle$, $|v\rangle$ can be determined

$$(a_0 + \zeta b_0)^2(1 - \cos\varphi) + c_0^2(1 - \cos\varphi) = 1 - (a_0 + b_0 + c_0)^2\cos\varphi \tag{2}$$

Once $|w\rangle$ is detected in the first stage, Bob tries to determine $|u\rangle$, $|v\rangle$ states in the second stage. Bob must get $|v\rangle$, because in our protocol Alice will send a number of photons in $|v\rangle$ polarization state which is equal to the number of impurities, i.e., $|v\rangle$ in $|u\rangle$ of the main text before the fence until he receives next $|w\rangle$. He sets his detector for $|v\rangle$, counts the number of detected photons and realizes the key with probability $b_0^2(1 - \cos\varphi)$. But, if $|w\rangle$ is not detected in the first stage photon will be in one of the polarization states between $|u\rangle$ and $|v\rangle$. For this, in the second stage of POVM he makes same kind of arrangement to identify the $|u\rangle$ and $|v\rangle$ polarization states of the photon. The corresponding measurement operators in the second stage of POVM will be:

$$A_u = \frac{P_u}{1 + S} \quad \text{where } P_u = (1 - |v\rangle\langle v|), \ S\langle u|v\rangle$$

$$\text{and } A_v = \frac{P_v}{1 + S} \quad \text{where } P_v = (1 - |u\rangle\langle u|).$$

If the angle between $|u\rangle$ and $|v\rangle$ be θ then $\langle u|v\rangle = \cos\theta$ and the detection of $|u\rangle$ results in $\langle u|A_u|u\rangle = (1 - \cos\theta)$ and the probability of getting ambiguous result is $\langle u|A_?|u\rangle = \cos\theta$.

The same is true for $|v\rangle$ polarized state and one can write for its detection as $\langle v|A_v|v\rangle = (1 - \cos\theta)$ and the probability of getting ambiguous result is $\langle v|A_?|v\rangle = \cos\theta$, while $\langle v|A_u|v\rangle = \langle u|A_v|u\rangle = 0$.

But, actually the photon is in combined polarization state $|\phi\rangle$, so the expectation value of detecting $|u\rangle$ is

$$\langle \phi|A_u|\phi\rangle = a_0^2(1 - \cos\theta) \text{ and } |v\rangle \text{ is} \langle \phi|A_v|\phi\rangle = b_0^2(1 - \cos\theta)$$

while the ambiguity is expected to be $\langle \phi|A_?|\phi\rangle = (a_0 + b_0)^2\cos\theta$.

The unitarity condition then demands that

$$a_0^2(1 - \cos\theta) + b_0^2(1 - \cos\theta) = 1 - (a_0 + b_0)^2\cos\theta \tag{3}$$

So far, we have analysed the detection criterion of polarization encoded photons in the undisturbed phase space of Bob's POVM. But this ideal situation may not occur in presence of eavesdropping or data loss during communication in the beam path, thereby introducing change of polarization states of photons. Then it is necessary to determine any kind of intervention and if it is within tolerable limit to proceed for key extraction by minimizing errors in communication. Eavesdropping may come from several possible ways as is mentioned earlier. In case of opaque eavesdropping, Eve measures each of the trits and then let it to proceed to Bob. Since any kind of quantum measurement changes the state of the photon so the photon states received by Bob usually will not be the one sent by Alice; because it should depend on the outcome of Eve's measurement. Eve's error rate in measurement will depend on the overlaps of the photon states

$$q = [\langle u|v\rangle + \langle u|w\rangle + \langle v|w\rangle]^2 = (\cos\theta + \sin\theta\sin\varphi + \cos\theta\sin\varphi)^2 \tag{4}$$

and the maximum information she can gather will be

$$I_{AE} = 1 + q\log_3 q + (1-q)\log_3(1-q) \tag{5}$$

In her stride for information she disturbs all the data and will easily be detectable. In order to conceal her presence she may intercept only a fraction, ξ of the data string. So she will loose information by the proportionate factor equal to the fraction of data intercepted. Bob in his two stage POVM will get some ambiguous results $R_1 = \langle w|A_?|w\rangle = \cos\varphi$ in first stage and $R_2 = \langle u|A_?|u\rangle = \cos\theta$ in the second stage. Therefore, after discarding the inconclusive results, Bob's error rate in the two stages will be determined by the amount of interception ξ and is given by

$$
\begin{aligned}
Qw &= \frac{q}{1-R_1} = \frac{\xi(\cos\theta + \sin\theta\,\sin\varphi + \cos\theta\,\sin\varphi)^2}{1 - \cos\varphi} \\
Qu,v &= \frac{q}{1-R_2} = \frac{\xi(\cos\theta + \sin\theta\,\sin\varphi + \cos\theta\,\sin\varphi)^2}{1 - \cos\theta}
\end{aligned}
\tag{6}
$$

For complete interception $\xi = 1$ and Bob's error rate Q is greater than q so no error correction or privacy amplification will be useful in this case and the communication has to be discarded.

On the other hand, not to reveal her presence, Eve may go for translucent eavesdropping where she also takes resort to POVM instead of direct measurement of photon states. With the probe photons having particular state of polarization, Eve slightly disturbs Bob's data interferometrically. The evolution of the polarization states of the photons results in change of mutual phase as well as the amount of overlaps among the photon states in the computational basis. Then the new equation for photon will be

$$|\psi'\rangle = e^{i\chi_1} a_0|u'\rangle + e^{i\chi_2} b_0|v'\rangle + c_0|w'\rangle \tag{7}$$

and the corresponding overlaps will be $\langle u'|v'\rangle = \cos\theta'$, $\langle v'|w'\rangle = \sin\theta'\sin\varphi'$ and $\langle u'|w'\rangle = \cos\theta'\sin\varphi'$ with $\theta' = \theta + \delta$ and $\varphi' = \varphi + \varepsilon$ accounts for the additional rotation of polarization angle by an amount δ and ε respectively. Now, Bob in his POVM while detecting the received photon states $|u'\rangle$, $|v'\rangle$ and $|w'\rangle$ should get different probabilities and the changed unitarity conditions for the two stages of POVM will be

$$(a_0 + \zeta'b_0)^2(1 - \cos\varphi') + c_0^2(1 - \cos\varphi') = 1 - (a_0 + b_0 + c_0)^2\cos\varphi' \tag{8}$$

and

$$a_0^2(1 - \cos\theta') + b_0^2(1 - \cos\theta') = 1 - (a_0 + b_0)^2\cos\theta' \tag{9}$$

This introduces an error in the estimation of the photon states as realized by Bob and can be determined by the projection of one photon states into the state space of the others. Then the error rates corresponding to the two stages of POVM will respectively be (the analysis for state v is same as state u)

$$q_w = \langle w'|A_\phi|w'\rangle = \frac{\sin^2\varepsilon}{1+\cos\varphi} \text{ and } q_u = \langle u'|A_v|u'\rangle = \frac{\sin^2\delta}{1+\cos\theta} \qquad (10)$$

While the probability of inconclusive results will be

$$R_1 = \langle w'|A_?|w'\rangle = \frac{\cos\varphi'}{1+\cos\varphi} \text{ and } R_2 = \langle u'|A_?|u'\rangle = \frac{\cos\theta'}{1+\cos\theta} \qquad (11)$$

After discarding the inconclusive results Bob will go for public discussion with Alice for estimation of his error rates corresponding to the two stages of POVM as

$$Qw = \frac{q_w}{1-R_1} = \frac{\sin^2\varepsilon}{\cos\varphi'} \text{ and } Qu = \frac{q_u}{1-R_2} = \frac{\sin^2\delta}{\cos\theta'} \qquad (12)$$

On forwarding the modified photons to Bob, Eve preserves her probe states to perform measurement of the probes at some later stage after hearing from Bob's measurement results. The information gain by Eve through this kind of intervention will be $I_{AE}(w) = 1 + q_w\log_2 q_w + (1 - q_w)\log_2(1 - q_w)$ on polarization state w and $I_{AE}(u) = 1 + q_u\log_2 q_u + (1 - q_u)\log_2(1 - q_u)$ for polarization state u or v. While the mutual information exchange in Eve-Bob channel will be determined by their joint estimation of errors as Eve is able to modify her error rates on hearing Bob of his error and will be $q_E(w) = Qw\cos^2\varepsilon + (1 - Qw)\sin^2\varepsilon$ corresponding to w polarization state and for u, v state it will be $q_E(u) = Qu\cos^2\delta + (1 - Qu)\sin^2\delta$. The estimation of information sharing $I_{BE}(w)$ and $I_{BE}(u)$ can be evaluated accordingly by using q_E. In this case in her stride extract more information Eve has increase the amount of intervention by increasing ε and δ and thereby eventually disturbing Bob's data, increases Bob's error and in turn looses share of information.

So, this is not the best strategy rather if Eve attaches her probe on the actual data through entangled translucent eavesdropping she can share more information from Bob's measurement. In doing so she will make a combined photon field where evaluation of the polarization state of a photon by Bob should depend on the detection of the state by Eve also. Then the error rate in the Alice-Bob channel before discarding inconclusive result will be the sum over all possible results obtained by Eve and can be written at the first stage of POVM as

$$q_{AB}(w) = \sum_{i=w,\phi} p(w, i, \phi) = c_0^2(1 - \cos\varphi')$$

while at the second stage of POVM

$$q_{AB}(u) = \sum_{i=u,v} p(u,i,v) = a_0^2(1 - \cos\theta') \quad \text{and}$$
$$q_{AB}(v) = \sum_{i=u,v} p(v,i,u) = b_0^2(1 - \cos\theta') \tag{13}$$

The rate of inconclusive result will be

$$R_w = \sum_{i=w,\phi} p(w,i,?) = (a_0 + b_0 + c_0)^2 \cos\varphi', \quad R_u = \sum_{i=u,v} p(u,i,?) = a_0^2 \cos\theta'$$
$$R_v = \sum_{i=u,v} p(v,i,?) = b_0^2 \cos\theta' \tag{14}$$

After discarding the inconclusive results the error rate in the Alice-Bob channel will be at the first stage

$$Q_{AB}(w) = \frac{q_{AB}(w)}{1 - R_w} = \frac{c_0^2(1 - \cos\varphi')}{1 - (a_0 + b_0 + c_0)^2 \cos\varphi'}$$

while at the second stage

$$Q_{AB}(u,v) = \frac{q_{AB}(u)}{1 - R_u} + \frac{q_{AB}(v)}{1 - R_v} = \frac{a_0^2(1 - \cos\theta')}{1 - a_0^2 \cos\theta'} + \frac{b_0^2(1 - \cos\theta')}{1 - b_0^2 \cos\theta'} \tag{15}$$

The information share in the Alice-Bob channel can be formulated in terms of $Q_{AB}(w)$ and $Q_{AB}(u,v)$ for w and u,v polarization states respectively as

$$I_{AB}(w) = 1 + Q_{AB}(w)\log_2 Q_{AB}(w) + (1 - Q_{AB}(w))\log_2(1 - Q_{AB}(w)) \quad \text{and}$$
$$I_{AB}(u,v) = 1 + Q_{AB}(u,v)\log_2 Q_{AB}(u,v) + (1 - Q_{AB}(u,v))\log_2(1 - Q_{AB}(u,v)) \tag{16}$$

In the same way one can determine the information share in the Alice-Eve and Bob-Eve channel through the estimation of error rate Q_{AE} and Q_{BE}.

For the Alice-Eve channel, the error rate is:

At the 1st stage

$$Q_{AE}(w) = q_{AE}(w) = \sum_{i=w,\phi,?} p(w,\phi,i) = c_0^2 \cos\varphi'$$

at the 2nd stage

$$Q_{AE}(u,v) = q_{AE}(u,v) = \sum_{i=u,v,?} p(u,v,i) = (a_0^2 + b_0^2) \cos\theta' \tag{17}$$

For Eve-Bob channel, the error rate will be

at the 1st stage

$$Q_{BE}(w) = \frac{q_{BE}(w)}{1 - R_w} = \frac{c_0^2(1 - \cos\varphi')\sin^2\varepsilon}{1 - (a_0 + b_0 + c_0)^2\cos\varphi'}$$

and at the 2nd stage

$$Q_{BE}(u, v) = \frac{q_{BE}(u)}{1 - R_u} + \frac{q_{BE}(v)}{1 - R_v} = \frac{a_0^2(1 - \cos\theta')\sin^2\delta}{1 - a_0^2\cos\theta'} + \frac{b_0^2(1 - \cos\theta')\sin^2\delta}{1 - b_0^2\cos\theta'} \quad (18)$$

By varying the amount of intervention i.e. by increasing the rotation angle of polarization ε and δ, Eve can increase the error rate in the Eve-Bob channel as is evident from the sine function at the numerator of the expression for error rate. She does this to hear from Bob about his measurement error and by doing so she improves her information share with Alice.

4 Conclusion

Experimental realization of qutrit [10] and entangled qutrit states [11, 12] have paved the way for its ever increasing use in quantum cryptographic key distribution schemes. A scheme for encryption of data in three non-orthogonal polarization states of a qutrit has been proposed in this paper. A detailed theoretical analysis has been made on the security aspects in presence of different eavesdropping strategies. An expression for the estimation of the maximum error rate and the rate of mutual information has been developed in terms of (i) the angle between the pair of non-orthogonal polarization states and (ii) the rate of transmission of photons. It is found that if the states are nearly orthogonal, i.e., φ and θ approaches $\frac{\pi}{2}$ even a small amount of eavesdropping invalidates the key distribution whereas the non-orthogonal states with small angle separation among the polarization states gives better result. In the interferometric scheme of POVM, this requires lower number of transmission of photon pulses. The analysis shows that information sharing in Alice-Eve channel is much lower in comparison to Eve-Bob channel irrespective of the eavesdropping strategies. This indicates that Bob's error will lead to Eve's information. The low intensity fencing state which encapsulates the location of key in the main data string increases the overall error rate in all the channels. Obviously, it limits the speed of key distribution, but this compromise has been made to improve security. In this article our intention is to propose a scheme of interferrometric measurement strategies for qutrit photon states and theoretical analysis of QBER and information reconciliation. It is true that we have not given any scheme of experimental implementation, but the work on this has been started since the first experimental realization of qubit in 2001in the hands of Knill, Laflamme and Milburn [13] followed by several such works by different groups; Zeilinger and coworkers [14–16], O'Brien and coworkers [17–20]. They have demonstrated single stage POVM and its variants for the determination of qubits.

Very recently when the scheme of utilization of entangled photons and qutrits or qudits have been theoretically proposed for QKD by Peres, Gisin and the coworkers [21] people have started thinking on its experimental realization. A few works have been reported [22] and different schemes and strategies for the minimization of the optical elements have been suggested. In this context our work on two-stage POVM is supposed to be implemented with fewer optical elements and we are in the process of estimating it.

References

1. Ekert, A.K., Huttner, B., Palma, G.M., Peres, A.: Phys. Rev. A **50**, 1047 (1994)
2. Bennett, C.H., Bassard, G.: Proceedings of IEEE International Conference on Computers, Systems and Signal Processing, Bangalore, India, p. 175. IEEE, New York (1984)
3. Brandt, H.E., Myers, J.M., Lomonaco, S.J.: Phys. Rev. A **56**, 4456 (1997)
4. Ekert, A.K., Rarity, J.G., Tapster, P.R., Palma, G.M.: Phys. Rev. Lett. **69**, 1293 (1992)
5. Wooters, W.K., Zurek, W.H.: Nature (London) **299**, 802 (1982)
6. Fuchs, C.A., Peres, A.: Phys. Rev. A **53**, 2038 (1996)
7. Fuchs, C.A., Gisin, N., Griffiths, R.B., Niu, C.S., Peres, A.: Phys. Rev. A **56**, 1163 (1997)
8. Peres, A.: Quantum Theory: Concepts and Methods (Kluwer, Dordrecht, 1993), Chap. 9
9. Myres, J.M., Brandt, H.E.: Meas. Sci. Technol. **8**, 1222 (1997)
10. Hong, C.K., Ou, Z.Y., Mandel, L.: Phys. Rev. Lett. **59**, 2044 (1987)
11. Bouwmeester, D., Pan, J.-W., Daniell, M., Weinfurter, H., Zeilinger, A.: Phys. Rev. Lett. **82**, 1345 (1999)
12. Eibl, M., Keissel, N., Bourennane, M., Kurtsiefer, C., Weinfurter, H.: Rev. Lett. **92**, 77901 (2004)
13. Knill, E., Laflamme, R., Milburn, G.J.: Nature (London) **409**, 46 (2001)
14. Schaeff, C., Polster, R., Lapkiewicz, R., Fickler, R., Ramelow, S., Zeilinger, A.: Opt. Exp. **20**, 16145 (2012)
15. Reck, M., Zeilinger, A., Bernstein, H.J., Bertani, P.: Phys. Rev. Lett. **73**, 58 (1994)
16. Zukowski, M., Zeilinger, A., Horne, M.A.: Phys. Rev. A **55**, 2564 (1997)
17. Li, H.W., Przeslak, S., Niskanen, A.O., Matthews, J.C.F., Politi, A., Shadbolt, P., Laing, A., Lobino, M., Thompson, M.G., O'Brien, J.L.: New J. Phys. **13**, 115009 (2011)
18. Laing, A., Peruzzo, A., Politi, A., Verde, M.R., Halder, M., Ralph, T.C., Thompson, M.G., O'Brien, J.L.: Appl. Phys. Lett. **97**, 211109 (2010)
19. Peruzzo, A., Laing, A., Politi, A., Rudolph, T., O'Brien, J.L.: Nat. Commun. **2**, 224 (2011)
20. Matthews, J.C.F., Politi, A., Stefanov, A., O'Brien, J.L.: Nat. Photon. **3**, 346 (2009)
21. Bechmann-Pasquinucci, H., Peres, A.: Phys. Rev. Lett. **85**, 3313 (2000)
22. Gelo Noel Tabia, M.: Phys. Rev. A **86**, 062107 (2012)

Evaluating the Performance of a Chaos Based Partial Image Encryption Scheme

Sukalyan Som, Sarbani Palit and Kashinath Dey

Abstract The traditional image encryption schemes, implementing fully layered encryption, seem to be undesirable in situations where there exists high data rate with limited bandwidth. A fully layered image encryption scheme first scrambles the image (to destroy the high correlation between neighbouring pixels) then encrypts it, resulting in a high computational overhead. In recent times, partial or selective encryption schemes are gaining popularity especially in situations demanding constrained communication such as mobile communications with limited computational power. In this paper, the performance of a chaos based partial image encryption scheme is evaluated in terms of a set of evaluation criteria. Exhaustive experimental simulation is performed to analyse the efficacy of the encryption scheme.

Keywords Partial image encryption · Tuneability · Visual degradation · Peak-Signal-to-Noise-Ratio (PSNR) · Mean-Square-Error (MSE) · Structural SIMilarity (SSIM) index

1 Introduction

As pointed out by Shannon [1], specific properties of images like high transmission rate and limited bandwidth, high redundancy and correlation make the use of standard cryptographic techniques inadequate. Thus a number of image encryption

S. Som (✉)
Department of Computer Science, Barrackpore Rastraguru Surendranath College,
Kolkata, India
e-mail: sukalyan.s@gmail.com

S. Palit
Computer Vision and Pattern Recognition Unit, Indian Statistical Institute, Kolkata, India
e-mail: sarbanip@isical.ac.in

K. Dey
Department of Computer Science and Engineering, University of Calcutta, Kolkata, India
e-mail: kndey@gmail.com

© Springer Nature Singapore Pte Ltd. 2017
R. Chaki et al. (eds.), *Advanced Computing and Systems for Security*,
Advances in Intelligent Systems and Computing 567,
DOI 10.1007/978-981-10-3409-1_12

173

algorithms are proposed where primarily confusion (scrambling) is performed to disturb the high correlation between adjacent pixels and thereafter diffusion (encryption) is achieved. Such schemes, referred to as *fully layered* or *total image encryption* schemes, seem to be unsuitable in situations where only limited resources v.i.z. high-definition delivery, low memory, low power are available. In recent times, a new approach known as *partial image encryption* has been explored where a subset of entire image data, chosen either statically or dynamically, is encrypted offering a trade-off between security and computational time. In this manuscript, we evaluate the performance of a chaotic tent map based biplane-wise partial encryption proposed in [2].

This paper is organized as follows: The partial image encryption scheme considered for the purpose of performance evaluation is briefly discussed in Sect. 2. Section 3 demonstrates the performance of the scheme in the light the of stated evaluation criteria with selected image databases. Conclusions are drawn in Sect. 4.

2 Overview of the Encryption Scheme

In this section we present a brief overview of the algorithm as described in [2] and now-onwards refer it as 'the algorithm'. The algorithm considers an 8 bits/ pixel gray scale image $I_{original} = (a_{ij})$ of size $2^n \times 2^n, n \in N$, a_{ij} representing the intensity of a pixel at the (i,j)th position. The entire algorithm consists of three sequential tasks— bitplane decomposition, assigning a threshold to determine significant bitplane and last of all, encryption of significant bitplanes through a chaos-based pseudo noise (PN) sequence generator to produce the cipher image.

2.1 Bitplane Decomposition

The pixel value a_{ij} is transformed into an 8 bit binary value and thus the image $I_{original}$ is converted into 8 binary images according to bit locations, known as bitplane images. In Fig. 1, 8 bitplane images of popular 'Lena' image is shown where (a) represents the LSB plane and (h) represents the MSB plane.

2.2 Assignment of Threshold to Determine Significant Bitplane

To determine the significance of a bitplane an autocorrelation based threshold is used. The determination procedure is summarized below:

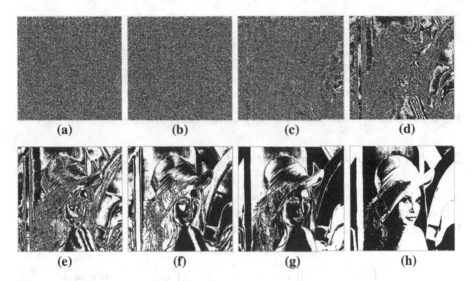

Fig. 1 Bitplane decomposition of original image 'Lena'

Step I: The original image $I_{original}$ is decomposed into 8 bitplanes. The kth bitplane image is denoted as $I^k(i,j), k = 0, 1, 2, \ldots, 7$.

Step II: Each bitplane is subdivided into four equal and disjoint blocks of size $2^{n-1} \times 2^{n-1}$. Each block is denoted by $B_m, m = 1, 2, 3, 4$ in row-major order.

Step III: For a block B_m, a vector b^m_{vec} is generated by stacking its columns in a single row so that its length is $2^{n-1} \times 2^{n-1}$. The Auto Correlation Function (ACF) of b^m_{vec} is considered for determining the significance of the bitplane. Let the elements of b^m_{vec} be denoted by $z_i, i = 1, \ldots, 2^{n-1} \times 2^{n-1}$ and their mean be $\bar{z}$. Then, the ACF coefficients for a lag τ, following [3], are given by,

$$r_\tau = \frac{c_\tau}{c_0} \tag{1}$$

where, $c_\tau = \frac{1}{N} \sum_{i=1}^{N-\tau}(z_i - \bar{z})(z_{i+\tau} - \bar{z})$

Step IV: A test statistic, t, defined below, is chosen to determine whether the kth, $k = 0, 1, 2, \ldots 7$, bitplane is significant or not.

$$t = max(r_\tau), \tau = -(L_B - 1), \ldots, -1, 1, 2, 3, \ldots, (L_B - 1) \tag{2}$$

r_τ is the autocorrelation of a vector B of length L_B at lag τ. At 5% level of significance the threshold is taken as 0.05. If $t \geq 0.05$ for a block, it is significant, leaving the remaining blocks for the bitplane to be tested. If $t < 0.05$, the block is insignificant and once all four blocks are identified, the bitplane is considered as insignificant.

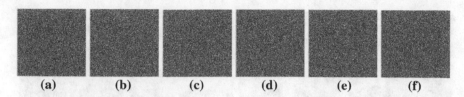

(a)   (b)   (c)   (d)   (e)   (f)

Fig. 2 Partial encryption of 'Lena': **a–f** Encrypted bitplanes of significant bitplanes of Fig. 1c–h

2.3 Encryption of Significant Bitplanes

A chaos based PN sequence generator, as proposed by [4], is used to generate bit sequence of length $2^n \times 2^n$. Comparing the outputs of the chaotic maps, given in Eq. 3, the bit sequence is generated as shown in Eq. 4.

$$x_{n+1} = \begin{cases} \mu x_n & \text{if } x_n < \frac{1}{2} \\ \mu(1-x_n) & \text{if } \frac{1}{2} \ge x_n \end{cases} \; and \; y_{n+1} = \begin{cases} \mu y_n & \text{if } y_n < \frac{1}{2} \\ \mu(1-y_n) & \text{if } \frac{1}{2} \ge y_n \end{cases} \quad (3)$$

$$g(x_{n+1}, y_{n+1}) = \begin{cases} 1 & \text{if } x_{n+1} \ge y_{n+1} \\ 0 & \text{if } x_{n+1} < y_{n+1} \end{cases} \quad (4)$$

The kth bitplane, I^k determined as significant is XORed bitwise with the generated PN sequence and the cipher image I_{cipher} is produced by concatenating the encrypted bitplanes with the unencrypted ones.

In Fig. 2, the encrypted and unencrypted bitplane images of 'Lena' is shown.

3 Performance Evaluation

A set of evaluation criteria—*Tuneability, Visual degradation, Cryptographic security, Encryption ratio, Compression friendliness, Format compliance, Error tolerance*, as proposed in [5] are commonly considered for the purpose of evaluating the efficacy of partial image encryption schemes. The encryption scheme summarized in Sect. 2 is evaluated on the basis of these evaluation criteria. For this purpose, along with different statistical tests, the algorithm is analysed against some common image processing attacks.

An extensive study of the algorithm has been performed using the USC-SIPI [6] and CVG [7] image databases which are collections of digitized images available and maintained by University of Southern California and University of Granada respectively. The algorithm, considered for performance evaluation, have been implemented using MATLAB 7.10.0.4(R2010a) on a system running with Windows 7 (32 bit) operating system with Intel Core i5 CPU and 4 GB DDR3 RAM.

Table 1 Average number of significant bitplanes per image database

Image database	Aerial	Textures	Sequences	Miscellaneous
No of significant bitplanes	5.00	5.08	6.67	6.32

Table 2 Number of significant bitplane-wise percentage (%) time savings (Avg.) per image database

Image database	Number of significant bitplanes								Average
	1	2	3	4	5	6	7	8	
Aerials	–	59.24	54.33	47.55	41.54	–	–	–	50.66
Textures	59.35	58.95	54.54	46.34	45.34	–	–	32.56	49.51
Miscellaneous	–	–	58.95	49.33	43.03	21.16	45.20	35.83	42.25
CVG	59.54	58.35	55.40	45.87	41.80	23.60	38.90	33.97	44.68

3.1 Tuneability

A partial encryption scheme must choose the target part for the encryption and encryption parameter dynamically. A static definition of the threshold severely restricts the usability the algorithm to a limited set of applications. The autocorrelation based threshold determines the significant bit planes which are then encrypted by the key-stream sequence generated by a cross-coupled chaotic tent map based bit generator. It has been found that a choice of 5% level of significance for the purpose of designing the threshold for significant bit plane determination provides a trade-off between the computational time and security aspects. In Table 1 the average number of significant bit planes for the images in USC-SIPI image database and CVG image database is shown. It may observed that the average number of significant bitplanes is not less than five and it ranges upto a maximum of almost seven.

The number of significant bit plane wise computational time saving for the gray images are shown in Table 2. It is to be noted that the table has several blank entries. It is an indication to the fact that there are databases where images with lower number of significant of bitplanes such as 1 and higher number of significant bitplanes i.e. more than 6 are not available. Another striking observation is that the time savings in case of images with 8 bitplanes significant is more than images with 6 bitplanes significant of the CVG database. This is because, a bitplane is determined to be significant by the autocorrelation based threshold from the first block only leaving the remaining three; whereas for images where some of the bitplanes are significant the ones that are non-significant are determined by computing the threshold values for all the four blocks.

Table 3 Visual degradation: average PSNR and SSIM per image database

Image databases	Aerial	Textures	Sequences	Miscellaneous
SSIM	0.0108	0.0093	0.0129	0.0094
PSNR	12.0453	11.0567	10.1212	10.12312

3.2 Visual Degradation

Visual degradation provides a measure of perceptual distortion of cipher image with respect to the original image. Some applications have the requirement of a high visual degradation while others may need quite the opposite. Static threshold based partial encryption schemes do not offer such variable amount of visual degradation. Being a tuneable partial encryption scheme the algorithm being evaluated, offers a varying level of visual degradation within the range of 10 dB to 12 dB using Peak-signal-to-noise ratio and 0.009 to 0.01 using structural similarity index (SSIM) respectively as metrics for the purpose. The results are shown in Table 3.

3.3 Cryptographic Security

One of the basic objectives of partial image encryption is to reduce the computational time yet offer a considerable level of security. As pointed out in [5] the cryptographic security depends on the encryption key and the unpredictability of the encrypted part. To prove the efficiency of the algorithm in terms of security three categories of tests v.i.z. based on image statistics, key sensitivity and key space and performance against attacks have been performed.

3.3.1 Tests Based on Image Statistics

1. **Correlation coefficient analysis** A secure partial image encryption technique must destroy the high correlation between the adjacent pixels within the image and at the same time the correlation between the original and cipher image must also be negligible. Pearson's product moment correlation coefficient, stated in Eq. 5, is used for the purpose of measuring the correlation.

$$r_{xy} = \frac{cov(x, y)}{\sigma_x \sigma_y} \; ; \sigma_x \neq 0 \; and \; \sigma_y \neq 0 \qquad (5)$$

In Table 4 average of correlation between the horizontally and vertically adjacent pixels of the original images and their cipher images are presented. Table 4 also presents the average of correlation between the original and cipher images of the databases. The drastic reduction in the correlation values in the cipher images as compared to that of the original ones is noteworthy.

Table 4 Correlation coefficient analysis

Image databases	Horizontally adjacent pixels		Vertically adjacent pixels		Between original and cipher image
	Original	Cipher	Original	Cipher	
Aerial	0.9798	0.0023	0.9844	0.0022	0.00018
Textures	0.8898	−0.0003	0.8946	−0.0042	0.00183
Sequences	0.9498	0.0043	0.9743	0.0012	−0.00115
Miscelleneous	0.9744	−0.0008	0.9632	−0.0004	−0.00053
CVG	0.9655	0.0012	0.9645	0.0042	0.00027

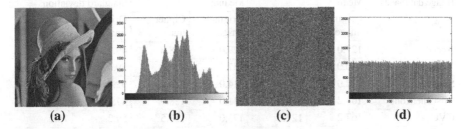

| (a) | (b) | (c) | (d) |

Fig. 3 Histogram analysis: **a** original Image 'Lena', **b** histogram of original image, **c** cipher image, **d** histogram of cipher image

2. **Test of homogeneity: Histogram analysis and χ^2 test** A secure cipher image must necessarily exhibit uniformity of distribution of the pixels implying a near uniform image histogram. The histograms of the 'Lena' and its corresponding cipher image are shown in Fig. 3 from which the inference can be made that the histogram of the original image shows a definite pattern and that of the cipher image is uniformly distributed. Histograms of all the images in chosen databases follows a similar behaviour.

The Chi-square test, as proposed in [8], is performed to assess the degree of uniformity of the distribution of the encrypted $M \times N$ image pixels analytically, which was shown in histogram analysis diagrammatically, as follows:

$$\chi^2 = \sum_{k=1}^{256} \frac{(v_k - \omega)^2}{256} \qquad (6)$$

where k is the number of gray levels (256), v_k is the observed occurrence frequencies of each gray level $(0 - 255)$, and $\omega = \frac{M \times N}{256}$. A 5% level of significance is considered for the computation of the tabulated values. The null hypothesis (H_0), the distribution of histogram of the encrypted image is uniform is accepted against the alternative hypothesis that the distribution of histogram of the encrypted image is non-uniform, if $\chi^2_{calculated} < (\chi^2_{256,0.05} = 293)$. In Table 5 the average val-

Table 5 Chi square value of images

Image name	$\chi^2_{calculated}$	$\chi^2_{tabulated}$
Aerial	244.2565	293
Textures	231.6457	293
Sequences	253.4343	293
Miscellaneous	256.2378	293
CVG	255.5544	293

Table 6 Measure of homogeneity: Mean, Median and Standard deviation in original and cipher image

Image databases	Mean		Median		Standard deviation	
	Original image	Cipher image	Original image	Cipher image	Original image	Cipher image
Aerial	127.0	128.1	129.0	128.3	49.2	75.0
Textures	140.0	128.5	137.6	127.8	59.6	74.3
Sequences	112.0	127.9	114.0	128.4	52.4	74.8
Miscelleneous	108.2	128.6	111.0	128.9	47.7	75.7
CVG	167.2	127.9	173.0	127.5	59.2	76.1

ues are shown which proves that the null hypothesis is accepted implying uniformity of pixels in cipher image.

3. **Measure of central tendency and dispersion** For a secure cryptosystem homogeneity of the cipher text symbols is highly sought. As measures of homogeneity, measures of central tendency through mean and median and measure of dispersion through standard deviation is adopted. The results for test of homogeneity is presented in Table 6. It can be inferred that that the average of measures in cipher images over the image databases are uniform i.e. cipher images have uniformity in mean, median and standard deviation irrespective of non-uniform values in original images.

4. **Information Entropy test** The entropy H(s) of a source signal s with $P(S_i)$ representing the probability of a symbol S_i, defined as

$$H(s) = \sum_{i=0}^{2^N-1} p(S_i) \cdot \log_2 \frac{1}{p(S_i)} \tag{7}$$

Assuming that the signal s emits 2^8 symbols with equal probability, the entropy $H(S) = 8$, corresponding to a truly random source s. The entropy of a cipher image should ideally be 8. Lesser the entropy, greater the chance of security threat. In Table 7 the average of entropy values for original images and their cipher images are presented. All cipher images are seen to have attained entropy values closer to the theoretically expected value 8 as compared to the original images.

Table 7 Measurement of encryption entropy

Image name	Entropy	
	Original image	Cipher image
Aerial	6.7326	7.9925
Textures	6.6728	7.9900
Sequences	6.8401	7.8312
Miscellaneous	5.9189	7.9948
CVG	6.2343	7.9896

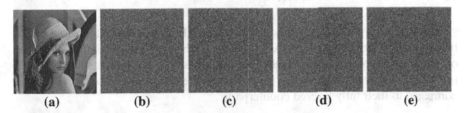

(a)	(b)	(c)	(d)	(e)

Fig. 4 Key sensitivity test: **a** original image (Lena 512×512), cipher image **b** with chosen secret keys $x_n = 0.101562$, $y_n = 0.101570$ and $\mu = 1.97$, **c** with change in only x_n (original: $x_n = 0.101562$, changed: $x_n = 0.101565$), **d** with change in only y_n (original: $y_n = 0.101570$, changed: $y_n = 0.101572$), **e** with change in only μ (original: $\mu = 1.97$, changed: $\mu = 1.97001$)

3.3.2 Key Sensitivity Analysis

A good encryption scheme should be sensitive to a minor change in the encryption key. Chaotic maps are becoming the popular choice for designing the partial image encryption schemes because of their intrinsic properties like sensitivity to initial conditions. To test the key sensitivity of the algorithm a tiny change $\Delta = 10^{-15}$ is applied to the key triplet (x_0, y_0, μ). In Fig. 4a original image of 'Lena' is shown where Fig. 4b represents its cipher with key $(x_0 = 0.101562, y_0 = 0.101570, \mu = 1.97)$. Figure 4c–e represents the cipher images with a minor change in each of the key keeping the others fixed. The correlation between the cipher images thus produced shows an average correlation of 0.0002 with the cipher image with the actual key.

3.3.3 Performance Against Attack: Differential Attack

As stated earlier, a desirable property of an image is sensitivity i.e. it should be sensitive enough to its key. This property is verified by changing one pixel of the original image. The cipher images thus produced must differ significantly. If a small change in the original image causes a significant change in the cipher image, then the differential attack is rendered inefficient. Two common measures of testing the influence of a one-pixel change on the whole image are used—Number of Pixels Change Rate (NPCR) and Unified Average Change in Intensity (UACI) were computed. The

NPCR of two cipher-images, C_1 and C_2 os size $M \times N$, which have only one pixel difference, is defined as

$$NPCR = \frac{\sum_{i,j} D(i,j)}{M \times N}, D(i,j) = \begin{cases} 0 & \text{if } C_1(i,j) = C_2(i,j) \\ 1 & \text{if } C_1(i,j) \neq C_2(i,j) \end{cases} \quad (8)$$

while the measure, UACI, is defined as

$$UACI = \frac{1}{M \times N} \sum_{i,j} \frac{|C_1(i,j) - C_2(i,j)|}{255} \times 100\% \quad (9)$$

For an efficient image encryption scheme the NPCR values must be close to 100 if expressed in percentage and UACI values must be near 33% [9, 10]. The average of NPCR values over all the images in the databases is 98.56% and that for the UACI is 31%. For a partial image encryption scheme the results are close to the ideal as compared to their fully layered counterparts.

3.4 Encryption Ratio

The ratio between the amount of encrypted part and entire data is termed as the *encryption ratio*. The objective of partial image encryption scheme is to minimize the encryption ratio and thereby reducing the encryption time so that it becomes a viable solution for high speed real time secure communication where the transmitter is having limited resources. Following [11], an image can be classified into three categories—Only a single bit plane containing the entire information, some of the bit planes containing image information and some other does not and all the bit planes containing image information. The partial image encryption scheme, being discussed here is confined to the first and second categories. Therefore the encryption ratio is always a proper fraction ($0 < encryptionratio < 1$) and consequently the amount of data being encrypted is always less than 100%. The higher is the encryption ratio, greater is the encryption time. In Table 8 the average of percentage of encryption being performed in all the image databases is presented. It can be observed that an average of 62–79% of image data is encrypted yielding an average time savings between 45 and 50%, as shown in Table 2.

Table 8 Encryption ratio: average of percentage of encryption performed in image databases

Image databases	Aerial	Textures	Sequences	Miscellaneous	CVG
Percentage of encryption	62.5	63.5	83.38	79.0	67.5

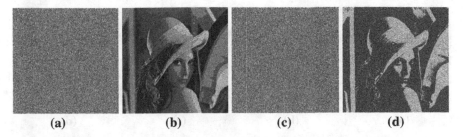

(a)	**(b)**	**(c)**	**(d)**

Fig. 5 Compression friendliness: **a** JPEG-compressed cipher image with QF 90, **b** decrypted image of (**a**), **c** JPEG-compressed cipher image with QF 20, **d** decrypted image of (**c**)

3.5 Compression Friendliness

A partial image encryption scheme is considered to be compression friendly if it has no impact or negligible impact on data compression. In Fig. 5 the encrypted image of 'Lena' compressed with quality factors 90 and 20 (shown in (a) and (c), respectively) along with their decrypted counterparts are shown in (b) and (d), respectively. The PSNR and SSIM values (average) between the decrypted images and the corresponding plain images are 32.7865 and 0.8756 for QF 90 and 29.0834 and 0.8452 for QF 20 respectively. It can be inference can be made that the decrypted images from the cipher images encountering JPEG compression are recognizable even for QF as low as 20.

3.6 Format Compliance

The encrypted image must be compliant with the compressor i.e. any standard decoder must be able to decode the encrypted data without decryption. This property is not being satisfied.

3.7 Error Tolerance

The partial image encryption scheme must be robust enough so that it can withstand noise attacks. Transmission of the cipher image is done through a public channel and thereby prone to undergo noise attacks. To evaluate the robustness of the algorithm the performance against two common noise-attacks, viz. salt and pepper noise and Gaussian noise have been verified. In Fig. 6 the encrypted images of 'Lena' attacked with Salt and Pepper noise of densities 0.005 and 0.05 (shown in (a) and (c), respectively)and their corresponding decrypted images are presented in (b) and (d), respectively.

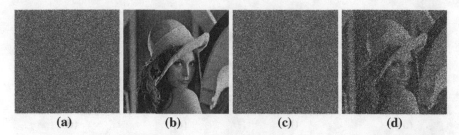

Fig. 6 Error tolerance: cipher image of 'Lena' attacked with Salt and Pepper noise with densities **a** 0.005, **b** 0.05, **c**, **d** decrypted images of (**a**), (**b**)

Fig. 7 Error tolerance: cipher image of 'Lena' attacked with AWGN with mean 0 and variance **a** 0.005, **c** 0.5, **b** decrypted image of (**a**), **d** decrypted images of (**c**)

In Fig. 7 the encrypted images of 'Lena' attacked with Additive White Gaussian noise (AWGN) with mean 0 and variances 0.005, and 0.5 (in (a) and (c), respectively) and their corresponding decrypted images are presented in (b) and (d), respectively. It is evident that it becomes increasingly difficult to decrypt the noise-attacked cipher image correctly with increasing levels of noise.

4 Conclusion

The performance of a simple, yet efficient chaos based partial image encryption scheme has been evaluated. An image can be considered as a combination of correlated and uncorrelated data where, most of the information is found to be present in the correlated part rather than the uncorrelated one. Hence, it would be sufficient to encrypt the correlated bitplanes instead of encrypting the entire image in order to speed up the overall procedure and save computational time. A threshold, based on the autocorrelation of disjoint block(s) of each bitplane, is used to determine the significant bitplanes of an image. Since most of the information in an image is found in the correlated part, it is sufficient to encrypt the correlated part leaving the insignificant part unencrypted. The significant bitplanes are encrypted using the PN sequences generated by chaotic maps. It has been noted that the choice of 5% level of significance based threshold makes the scheme a tuneable one offering varying levels of visual degradation which is a desirable feature. The scheme offers around 45% savings in time yet provides high security.

References

1. Shannon, C.E.: Communication theory of secrecy systems, Declassified Report (1946)
2. Som, S., Kotal, A., Mitra, A., Palit, S., Chaudhuri, B.B.: A chaos based partial image encryption scheme. In: Proceedings of IEEE International Conference on ICBIM 2014 (2014)
3. Box, G.E.P., Jenkins, G.M., Reinsel, G.C.: Time Series Analysis: Forecasting and Control, 3rd edn. Prentice-Hall, Upper Saddle River, NJ (1994)
4. Pareek, N.K., Patidar, V., Sud, K.K.: A random bit generator using chaotic maps. Int. J. Netw. Secur. **10**(1), 32–38 (2010)
5. Massoudi, A., Lefebvre, F., Vleeschouwer, C., Macq, B., Quisquarter, J.: Overview on selective encryption of image and video: challenges and perspectives. EURASIP J. Inf. Secur. (2008)
6. University of Southern California, Signal and Image Processing Laboratory. http://sipi.usc.edu/database. Accessed 12 Jan 2013
7. Computer Vision Group, University of Granada. http://decsai.ugr.es/cvg. Accessed 12 Jan 2013
8. Jolfaei, A., Mirghadri, A.: Image encryption using chaos and block cipher. Comput. Inf. Sci. **4**(1), 172–185 (2011)
9. Menezes, A.J., Oorschot, P.C.V., Vanstone, S.A.: Handbook of Applied Cryptography. CRC Press, Boca Raton (1997)
10. Chen, G., Mao, Y., Chui, C.K.: A symmetric encryption scheme based on 3D chaotic map. Chaos Solitons Fract. **21**, 749–761 (2004)
11. Mitra, A., Palit, S., Chaudhuri, B.B., Kundu, S., Pathak, S., Datta, R.: A new partial image encryption method for secure multimedia communication. In: Proceedings of Workshop on Mobile Systems (WoMS), WBUT, Kolkata (2008)

Keystroke Dynamics and Face Image Fusion as a Method of Identification Accuracy Improvement

Piotr Panasiuk, Marcin Dąbrowski and Khalid Saeed

Abstract This paper concerns about keystroke dynamics and face image fusion. Different methods of database collecting are presented. The authors combined data from their own keystroke database and public AT&T *Database of Faces*. *Keystroke Dynamics—Benchmark Data Set* was used additionally. Two selected approaches have been merged and overall system reliability was tested. Initial outcome shows that the results when the algorithms work together give more robust identification accuracy.

Keywords Keystroke dynamics · Face image · Identification · Behavioral biometrics · Physiological biometrics · Database acquisition · Computer security · Fusion · Multimodal biometric system

1 Introduction

In this paper authors are trying to combine two biometric features and use them together to achieve improvement in system's overall user identification accuracy. The said features are keystroke dynamics and face image. Both of these features provide quite reliable results when used separately, but it may occur that error rates are not low enough for the system to be used in applications where unconditional security with no place for mistakes is really demanded. Such use may be considered when using ATMs or in authenticating bank transactions. Another perfect example

P. Panasiuk · K. Saeed (✉)
Faculty of Mathematics and Information Science, Warsaw University of Technology, Warsaw, Poland
e-mail: k.saeed@mini.pw.edu.pl

P. Panasiuk
e-mail: p.panasiuk@mini.pw.edu.pl

M. Dąbrowski · K. Saeed
Faculty of Computer Science, Bialystok University of Technology, Bialystok, Poland
e-mail: marcin.dabrowski@poczta.fm

© Springer Nature Singapore Pte Ltd. 2017   187
R. Chaki et al. (eds.), *Advanced Computing and Systems for Security*,
Advances in Intelligent Systems and Computing 567,
DOI 10.1007/978-981-10-3409-1_13

could be a healthcare system where data security and privacy is a vital feature. General lack of such systems in widespread use can lead to assumption that they are not developed enough to provide satisfying results. Such fusion approach is really needed where single biometric feature can be prone to errors or ease of imposture.

2 Known Approaches

Both face image recognition and keystroke dynamics are very popular among all biometric methods. There have been many approaches done in both of these domains.

The most recent works describes noise- and illumination-tolerant face recognition, where authors propose band-pass correlation filter [1]. Using test face images corrupted by noise, authors obtained recognition improvement over standard low-pass and high-pass filters. In [2] authors combined keystroke dynamics with soft biometrics. Gender, age and handedness were considered. The best EER value of 5.41% was obtained.

Authors of this paper presented multiple experiments in the field of biometrics in their previous works, mostly concerning keystroke dynamics. These include algorithms using average keystrokes per minute, combined and separated dwell and flight times, typing errors, key overlapping and many others. One of notable works includes obtaining classification accuracy of 90.83% for 21 users, where samples contained longer sentences and short phrases in both Polish and English [3]. Algorithms from previous experiments were later modified for use in next works. This resulted in accuracy of 98.78% in experiment based on fixed-text, where database consisting of 16 users was used. However higher user count resulted in decreased accuracy (e.g. 72.3% for 79 persons) [4]. Another work by the authors examines keystroke database quality [5]. It was proven that databases consisting of longer keystroke samples are more appropriate for user identification than authentication. What is worth noting, keeping imperfect user-specific samples in database lowered FRR. In the next work [6] authors experimented with authentication using different length non-fixed text. Users samples were collected locally and over the web using suitable applications developed by the authors. Results showed higher recognition accuracy for longer texts and lower for shorter ones respectively. When statistical characteristics of the sample were considered, better results were obtained in comparison to using raw sample data. Finally EER of 6.1% was obtained when using samples consisting of 200 keystrokes. In [7] authors collected their own database for comparison with *Keystroke Dynamics—Benchmark Data Set* [8]. It was almost identical to compared database, but authors collected it in less restrictive conditions. In contrary to [8] it was unsupervised and commonly available equipment was used. Same algorithms ran on both databases gave even 30% advantage for authors approach in some situations. As less restrictive way of collecting data resulted in better outcome, authors believe that testing new algorithms on databases from multiple sources may be a good practice.

3 Sample Data Collecting

Keystroke dynamics samples were collected from multiple users using a few methods. Some of these methods can be described as follows.

3.1 Keystroke Data Collection (Author's Initial Test Database)

The data collecting application was written in C#. At first users selected their own usernames and passwords, to be able to use system freely multiple times. Using their own hardware, eight users were typing three authors' selected ~200–300 character long texts. Two of them were fragments of different novels in Polish and English respectively and the last text consisted of pangrams in both languages. The texts were chosen in such a way, that every letter in the 26-letter English alphabet was used at least once in three texts combined. These selected texts were typed one after another in a single session. There were two such sessions for each user conducted altogether with at least one day break between them. One of the users was excluded, because of not completing two sessions.

Since it was difficult to gather greater number of individuals willing to contribute much more samples to authors' database, publicly available *Keystroke Dynamics—Benchmark Data Set* [8] was examined additionally.

3.2 AT&T Database Collection

For face image samples, publicly available AT&T database [9] consisting of 40 users was used. The reason was to get results comparable with other published approaches. The data in AT&T database was collected in a way that each one of 40 users has 10 face images taken in various conditions (changing lighting, facial expressions and details—person wearing glasses etc.). Images were normalized to eyes line (similar position in each sample), cropped and converted to grayscale to reduce number of dimensions.

3.3 Keystroke Dynamics—Benchmark Data Set

Database built by Kevin Killourhy and Roy Maxion consists of 51 users and 400 short keystroke samples for each of them. All users typed the same phrase—". tie5Roanl". Dwell and flight times were gathered. Data were collected in 8 sessions of 50 correctly typed samples. Said experiment was supervised and specific high-precision devices were used to get high quality data. Authors of this paper used described database because of its public availability and large collection of quality samples from many individuals.

4 Classification

Two methods are presented, as authors of this paper found it necessary for the experimental setup to be redeveloped during research.

4.1 Authors' Initial Approach

The presented method proposition consists of using nearest neighbor classifier with Euclidean metric for keystroke dynamics and Eigenfaces algorithm for face recognition. Fisherfaces algorithm was also used for comparison.

In the first step, user keystroke data are examined. Only key dwell times are taken into consideration. Keystroke samples collected during the first session in form of three texts are being averaged separately for each user, making corresponding user training vectors. Similarly data from the second session are used to create test vectors. Euclidean distance is calculated to define affiliation to classes in testing phase of experiment.

Secondly, user face image is analyzed. Training set is built from given number (from 1 to 9 per user—same for all users) of randomly chosen samples. Testing set consists of remaining face images which were not used for training. Then OpenCV is used to perform classification. That is nearest neighbor using Euclidean metric.

If results of both tests are positive, user gains access. Otherwise, if none of tests passes, user is being rejected.

It was noticed that users whose native language is Polish, were typing Early Modern English text with more difficulty than Present-Day English or Polish texts. In case of face recognition, lighting was found to be very important factor affecting the classification. This is characteristic for Eigenfaces algorithm [10].

For the data to be able to be used together, the face image database had to be truncated to seven users—same amount as in the authors' keystroke database. Seven first users from AT&T database were chosen. Face images were assigned to succeeding users in keystroke database.

During the experiment it was noticed that small number of user samples led to producing highly unreliable results. In this case authors chose to abandon further experiments using this method. It was later redeveloped in a new way to accept samples from *Keystroke Dynamics—Benchmark Data Set* among other things. Despite this, authors preset outcome from both experimental setups in results section.

4.2 Second Approach

This time authors decided to combine AT&T *Database of Faces* with *Keystroke Dynamics—Benchmark Data Set*. It is believed to result in reliable outcome.

Face database includes 10 samples for each of 40 users and keystroke database consists of 400 samples for each of 51 users. Because both databases consisted of different number of users, authors chose to trim keystroke database to 40 users, therefore it matches face database. No test samples of users which data does not exist in databases were used.

Photos carry much more data than keystroke samples and because of that for each one of training face samples 10 keystroke samples were assigned from corresponding users in keystroke database. To simplify the setup and increase performance only dwell times were taken into consideration when it comes to keystroke samples. In this setup Eigenfaces algorithm was used for face images as authors found it fast and reliable. In this experiment k-Nearest Neighbors algorithm and its weighted variant was used for classification. Distances were calculated using Euclidean metric.

In the beginning all samples with corresponding user ID's from keystroke and face databases are loaded in exact way they are stored. Next user training and testing samples numbers are determined. Authors analyzed from 1 to 9 training samples per user for face database and from 10 to 90 training samples from keystroke database accordingly. Testing samples consisted of remaining samples which were not used for training. For example, when 4 face samples were training samples, 6 samples were used for testing. In case of keystroke samples same number of testing samples was used as from face database. For example, when 40 keystroke training samples were used, 6 samples were used for testing, due to being strictly connected to face samples. This represents real world situation when user is typing password one time and his photo is also taken once.

When the number of training and testing samples is determined, they are chosen randomly from both databases, while keeping in mind users to which they originally belonged. This is true for over 7000 algorithm passes, where in one pass up to 360 testing samples combined of image and keystroke data are being analyzed. User ID affiliation to both databases does not change, but samples are chosen randomly for each algorithm pass.

In the next step, distances for testing and training samples are calculated using Euclidean metric. Face data is analyzed using Eigenfaces algorithm. All distances with corresponding classes are stored to be used in k-NN classification. The value of k from 1 to 9 was tested. The fusion relies on weighted k-NN algorithm, where different weights are assigned to keystroke and face data accordingly. Weight values from 0.1 to 0.9 in steps of 0.1 were tested for both characteristics. The sum of both weights is always equal 1. For each of testing samples, k minimal distances with corresponding classes are saved and weights are applied. The sum of weighted values representing class affiliation from keystroke and face data determines final class to which current testing sample will be assigned.

The results are stored in a way it is possible to tell if samples were identified correctly. Identification accuracy for current pass is noted and experiment is repeated for total of 10 times with samples chosen randomly for each pass. These results are averaged to minimize errors. After 10 passes of the algorithm, different number of training and testing samples, different k or different weights are chosen to be examined next.

5   Results

It has been observed that *Keystroke Dynamics—Benchmark Data Set* gave more reliable results than authors' database. Therefore fusion was only performed on AT&T *Database of Faces* and said keystroke database. Authors' database was excluded from further examination due to its small size. Following results from both approaches were obtained in this research.

5.1   Initial Approach

Regarding keystroke dynamics the sample consisting of three text phrases from the first session were used as a training set and samples from the second session were used for testing. Using such small sets allowed to obtain user identification accuracy of 71.4%, i.e. 5 out of 7 users were recognized correctly. According to these results it can be stated that using only few training samples for each user is not enough to get satisfying outcome. Results may differ greatly with database containing much more users.

When it comes to face recognition using Eigenfaces, the training set contains a single sample for each of 7 users and the testing set consists of 9 images for each user. Accuracy ranging from 76.2 to 95.2% was obtained for different selections of 7 users from database of 40 users. Experiment using whole database of 40 users and same training and testing set distribution resulted in accuracy of 61%. All of above results were obtained whilst using 10 Eigenfaces for training (Fig. 1).

Just one training sample for each user allowed to give us high recognition success ratio for small user sets. Decreased accuracy has been observed for database containing more users. When larger number of users is in regard, a chance of their face likeness is higher, thus making some of them harder to identify.

In case of Fisherfaces training and testing sets with ratios from 1:9 to 9:1 accordingly were used. This is similar to previous experiment with Eigenfaces. Different selections of 7 users from database of 40 users resulted in identification accuracy ranging from 63.1 to 90.4%. For the whole database of 40 users accuracy

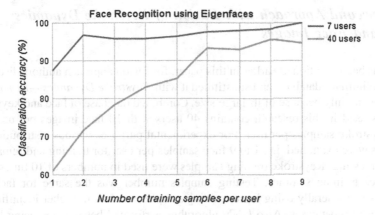

Fig. 1 Eigenface classification accuracy for databases of different number of users with varying training and testing set sizes

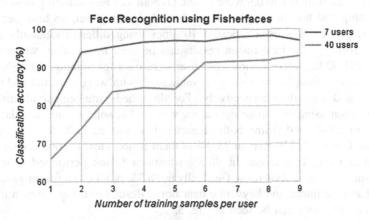

Fig. 2 Fisherface classification accuracy for databases of different number of users with varying training and testing set sizes

of 66% has been obtained. All available Fisherfaces were used for classification. Presented results are true for single training sample and 9 testing samples per user. Results for different set sizes are shown additionally in Fig. 2.

Fisherface classification resulted in higher accuracy than Eigenface classification when using database of 40 users with low number of training samples. However, Eigenfaces gave better outcome than Fisherfaces for smaller database of 7 users also with low number of training samples.

Overall Eigenfaces algorithm performed slightly better than Fisherfaces algorithm, when bigger training sets were used. This may be due to specific characteristics of used database [9]. Authors decided to use this database in their fusion approach.

5.2 Second Approach with Fusion of Keystroke Dynamics and Face Images

As it has been mentioned earlier in this paper, for fusion approach authors discarded their preliminary database and substituted it with *Keystroke Dynamics—Benchmark Data Set*, mainly because of its large size. Combined database of face and keystroke samples used in this research contains 40 users with 10 face images per user and 400 keystroke samples per user. For experimental purposes various distributions of samples were examined, i.e. 1 to 9 face samples per user for training and remaining ones for testing. Keystroke training samples were used in numbers of 10 times more than face training samples. Testing samples number was the same for face and keystrokes. Generally using more training samples resulted in higher identification accuracy in most cases. Also k-NN algorithm performed better when using higher k values with more training samples accordingly.

Authors' fusion algorithm uses different weights for face and keystroke samples. Values ranging from 0.1 to 0.9 were tested. Overall 729 tests differing in number of used training and testing samples, k values and changing weights have been conducted. Every test has been repeated 10 times using different randomly chosen samples. Some of the best fusion results can be seen in situation, where 6 face samples and 60 keystroke samples per user were used for training, 4 face samples and 4 keystroke samples for testing, k value of 3, with weights of 0.4 and 0.6 for keystroke and face data respectively. For described arrangement identification accuracy when using keystroke data alone was 68.4%, solely using face data gave accuracy of 85.4% and fusing both characteristics resulted in 93.5% correct identifications. Out of all 729 tests in 78.9% of them fusion identification accuracy was higher than using face alone. In 98.5% situations fusion performed better than solely using keystroke dynamics. Generally in 78.3% of all cases fusion gave better results than face image and keystroke data alone. Results of using different training and testing sets sizes can be seen in Fig. 3.

It can be noticed that for above configuration identification accuracy is always greater when using fusion method. Low accuracy for little numbers of training samples when using only face for identification can be explained by nature of k-Nearest Neighbor classifier. In this case $k = 3$ was used which is too high value for 1 or 2 training samples. Because of that 1-NN classifier sometimes performed better than higher k values. This isn't the case for keystroke data, because these training samples were used in multiples of 10. Identification accuracy by using keystroke data is staying on similar level most of the time because not much is gained by using more than 10 keystroke samples. Keystroke dynamics being behavioral biometric feature is better suited for authentication, but can be very good supplement for physiological features like face image in this research.

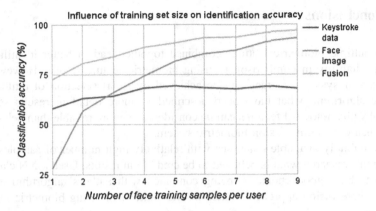

Fig. 3 Identification accuracy for training samples ranging from 1 to 9 for face, 10 to 90 for keystrokes, $k = 3$ and weights of 0.4 and 0.6 for keystrokes and faces respectively

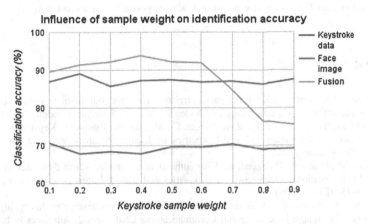

Fig. 4 Identification accuracy for 7 training samples for face, 70 for keystrokes, $k = 3$ and weights ranging from 0.1 to 0.9 for keystroke samples

In Fig. 4 results are shown for different weights used for keystroke and image data.

Fusion results depend mainly on weights applied to keystroke and face data samples. It can be seen that using lower weights for keystrokes may increase identification accuracy in fusion approach. Sum of weights for both characteristics is always equal 1. Face image is generally more reliable biometric feature than keystroke dynamics and when face samples weight is lower or equal to keystroke data weight sharp decrease of fusion classification accuracy can be observed.

Overall identification accuracy is higher when face image is considered as main feature and keystroke dynamics is taken as an additional feature.

6 Conclusions

Using multiple biometric features combined together lead to better identification system reliability in comparison to using biometric features alone. However the vital gain of system accuracy can be achieved by smart selection of multimodal balance algorithm, what have been described in more detail in results section. Generally physiological features can be considered as more reliable than behavioral ones when developing fusion biometric system.

Two publicly available databases with relatively high number of samples were used in this research what is believed to be credible in results. Used k-NN classifier proved to be a good choice as being combined with authors' algorithm it gave results where fusion approach's outcome was better than using biometric features individually.

Acknowledgements This work was partially supported by grant number S/WI/1/2013 from Bialystok University of Technology and funded from the resources for research by Ministry of Science and Higher Education. It was also partially supported by Neitec company.

References

1. Banerjee, P.K., Datta, A.K.: Band-pass correlation filter for illumination- and noise-tolerant face recognition. In: Signal, Image and Video Processing. Springer, London (2016)
2. Syed Idrus, S.Z., Cherrier, E., Rosenberger, C., Mondal, S., Bours, P.: Keystroke dynamics performance enhancement with soft biometrics. In: Identity, Security and Behavior Analysis. IEEE, Hong Kong (2015)
3. Rybnik, M., Panasiuk, P., Saeed, K.: User authentication with keystroke dynamics using fixed text. In: International Conference on Biometrics and Kansei Engineering, Cieszyn, Poland, pp. 70–75. IEEE (2009)
4. Panasiuk, P., Saeed, K.: A modified algorithm for user identification by his typing on the keyboard. In: Image Processing and Communications Challenges 2. Advances in Intelligent and Soft Computing, vol. 84, pp. 113–120. Springer, Heidelberg (2010)
5. Rybnik, M., Panasiuk, P., Saeed, K., Rogowski, M.: Advances in the keystroke dynamics: the practical impact of database quality. In: Computer Information Systems and Industrial Management. Lecture Notes in Computer Science, vol. 7564, pp. 203–214. Springer, Berlin (2012)
6. Rybnik, M., Tabedzki, M., Adamski, M., Saeed, K.: An exploration of keystroke dynamics authentication using non-fixed text of various length. In: International Conference on Biometrics and Kansei Engineering, pp. 245–250. IEEE (2013)
7. Panasiuk, P., Dąbrowski, M., Saeed, K., Bocheńska-Włostowska, K.: On the comparison of the keystroke dynamics databases. In: Computer Information Systems and Industrial Management. Lecture Notes in Computer Science, vol. 8838, pp. 122–129. Springer, Berlin (2014)
8. Killourhy, K.S., Maxion, R.A.: Comparing anomaly-detection algorithms for keystroke dynamics. In: Dependable Systems & Networks, Lisbon, Portugal, pp. 125–134. IEEE (2009)
9. AT&T Database of Faces: http://www.cl.cam.ac.uk/research/dtg/attarchive/facedatabase.html. Accessed 30 Apr 2016
10. Turk, M., Pentland, A.: Eigenfaces for recognition. J. Cogn. Neurosci. **3**(1), 71–86 (1991)

Author Index

© Springer Nature Singapore Pte Ltd. 2017   197
R. Chaki et al. (eds.), *Advanced Computing and Systems for Security*,
Advances in Intelligent Systems and Computing 567,
DOI 10.1007/978-981-10-3409-1

Printed in the United States
By Bookmasters